CHEMICAL MANAGEMENT

CHEMICAL MANAGEMENT

Reducing Waste and Cost Through Innovative Supply Strategies

THOMAS J. BIERMA, MBA, PHD

AND

FRANCIS L. WATERSTRAAT, JR., MBA

Illinois State University

John Wiley & Sons, Inc.

New York • Chichester • Weinheim • Brisbane • Singapore • Toronto

This publication is designed to provide accurate and authoritative information in regard to the subject matter covered. It is sold with the understanding that the publisher is not engaged in rendering professional services. If professional advice or other expert assistance is required, the services of a competent professional person should be sought.

Library of Congress Cataloging-in-Publication Data:
Bierma, Thomas J.
  Chemical management : reducing waste and cost through innovative
supply strategies / Thomas J. Bierma and Frank L. Waterstraat.
      p.   cm.
  Includes bibliographical reference and index.
  ISBN 0-471-33284-4 (alk. paper)
  ISBN 13: 978-0-471-33284-8
  1. Industrial ecology. 2. Waste minimization. 3. Chemicals—
Management. I. Waterstraat, Frank L. I. Title.
TS161.B56   2000
658.5—dc21   00-16024

10 9 8 7 6 5 4 3 2 1

To JoAnne, Alissa, Wesley, Kathy, Kim, Kirsten, and Kelly—
for your patience, support, and understanding

CONTENTS

PART 2   SHARED SAVINGS CHEMICAL MANAGEMENT

PREFACE

Have you ever heard someone's opinion and thought, "That is absolutely crazy," but long afterward, looking back, it didn't seem so crazy after all? In fact, it made so much sense you wondered how you could ever have missed it!

You may have the same experience with Shared Savings chemical management, the innovative chemical supply strategy introduced in this book. We certainly did! Our initial reaction was that it simply can't work. But after four years of research into companies with successful Shared Savings programs, including the five plants featured in this book, we know that it not only works, it's hard to imagine how companies can continue to manage their chemicals the old way.

We first learned of Shared Savings through a story told to us about a Ford Motor Company plant in Canada. At this plant, the chemical supplier was behaving in a very strange manner. The plant's supplier was actually working on a daily basis with plant personnel to *reduce* the amount of chemicals the plant needed from the supplier to produce cars! We chalked it up to another big auto company squeezing another poor supplier to the point of delirium.

Later, we learned that suppliers were behaving in a similar manner at many General Motors plants as well. Finally, when we had the opportunity to interview some of these suppliers, to our amazement we found they liked their relationship with the auto manufacturers! They even preferred it over the old sales relationship! In fact, the people we interviewed from both suppliers and manufacturers wouldn't have it any other way! They would never go back to the old way of doing business.

The reason for their happiness is this: traditional supply relationships are inherently—and extremely—wasteful. The new Shared Savings programs are structured to turn that waste into profit. That the profit is *shared* by both manufacturer and supplier drives the relationship to ever greater levels of performance.

The dilemma is perhaps best expressed in two of our favorite quotations from the many interviews we conducted in our research. Jerry Mittlestaedt of Navistar International, Inc., explained:

> It saves us tons of money and improves the environment. This whole thing is just good business.

But, warned Chrysler's Bob Conrad:

> It requires a whole different kind of thinking.

And so it does. We have spoken to hundreds of businesspeople around the nation about Shared Savings and how it works in the companies we studied. There are two common reactions:

1. **You must be wrong.** Shared Savings can't possibly work. Just give it a little more time and the whole thing will fall apart. Well, some companies have used Shared Savings for over a decade. The five plants we profile in this book have over thirty-five years of combined Shared Savings experience.
2. **These companies are unique.** It will never work in other businesses. In our experience, the most unusual thing about the companies we studied is that they had been *desperate.* They had all gone through times that were so bad (the U.S. automotive, farm equipment, and railroad equipment crises of the 1980s) that they were *forced* to change. Ideas that seemed outrageous before were finally given a chance.

But in every group we speak to, a few individuals react differently. They ask:

> How can I make it work for my company?

It is for these individuals that we have written this book.

We are asking you not to dismiss Shared Savings with "That is absolutely crazy." We believe that as you learn more about Shared Savings, you will wonder why the idea took so long to develop.

We have written this book primarily for managers—managers of Production, Purchasing, Maintenance, Environmental Control, Health and Safety, Engineering, or any of the multitude of departments touched by chemicals in today's manufacturing plants. Though we write of chemicals used in creating products, the concepts apply equally as well to the creation of services, a point that is illustrated by the innovative chemical management program initiated at Delta Air Lines (see chapter 14).

Following the introductory chapter, the book is divided into four parts. Part 1 introduces the Chemical Beast—that never-ending cycle of headaches and escalating costs associated with traditional chemical supply programs. Part 2 examines alternatives to the traditional approach to chemical supply and introduces Shared Savings chemical management. In Part 3, we provide case histories of plants with successful Shared Savings programs, including in-depth profiles of five U.S. plants. Finally, Part 4 explores the process of implementing a Shared Savings program.

We are grateful to the many companies and employees who so generously donated their time to our education. We also wish to recognize the Chemical Strategies Partnership of San Francisco, the U.S. Environmental Protection Agency, and the Illinois Waste Management and Research Center for their efforts to make chemical management an effective environmental policy tool. We would especially like to thank the many individuals who reviewed early drafts of this book and provided us with such thoughtful and constructive comments. We are indebted to the staff at John Wiley & Sons, Inc. for their efforts in making this book a reality. In particular, we would like to thank Neil Levine, Rose Kish, and Maury Botton, with a special thanks to Madeline Perri for her diligent work on the manuscript.

Of course, in a highly competitive market, things change quickly. People change jobs, programs evolve, even companies change. Since we completed our interviews for this book, Chrysler has become DaimlerChrysler, and BetzDearborn has been acquired by Hercules, Inc. We have not reflected these changes in our book, since the programs we studied were those of Chrysler and BetzDearborn, respectively. Only time will tell what the programs of new companies will be.

Shared Savings is certainly "a whole different kind of thinking." If we hadn't seen it (more than once!) we never would have believed it. Now we can't imagine a chemical supply relationship structured any other way!

Thomas J. Bierma and Francis L. Waterstraat
Normal, Illinois
September 1999

It's Time to Change Your Chemical Management Strategy

It saves us tons of money and improves the environment.
This whole thing is just good business.
—JERRY MITTLESTAEDT,
Environmental Control Manager, Navistar International, Inc.

Imagine a different world, a world where *your* chemical supplier

- continuously searches for ways to *minimize* the amount of chemicals you need in order to produce a quality product or service
- helps improve your operations in order to *reduce* the amount of chemicals that become waste
- continuously finds opportunities to *increase* the *quality* of your products and the *value* you can offer the ultimate consumer
- provides you with additional chemical and environmental management *resources* and increases chemical and environmental *performance* while delivering *cost savings* to plant management
- collects data regularly on chemical usage and inventory and provides it to you as needed for *environmental reporting* and the *efficient management of chemicals*

Sound like a dreamworld? It's a reality for some managers in manufacturing today. Their companies have adopted an innovative chemical supply strategy that we call *Shared Savings Chemical Management*. It is a simple strategy—suppliers are given a direct financial stake in chemical performance and efficiency. It not only provides key benefits to management, such as reduced operating costs and improved production, but also greatly improves environmental performance. However, as one chemical manager put it, "It takes a whole different kind of thinking."

Many companies use chemicals in their processes to make products or provide services. These may range from chemicals that become part of the product, such as paints and plastics, to chemicals used in various processes, such as oils and cleaners. Yet with the use of chemicals come problems, headaches, and costs. Consider the costs of chemical purchasing, managing inventory, regulatory compliance, waste disposal, and numerous other chemical-related activities. These costs are not small—in fact, they may be several times greater than the purchase price of the chemicals themselves— and they can occur despite the best intentions and efforts of both chemical user and chemical supplier.

Our research has found that some companies are dramatically reducing their total chemical costs and improving chemical performance by abandoning the traditional chemical supply relationship. Instead, they are implementing innovative approaches to align supplier financial interests with their own, substantially reducing the number and volume of chemicals used.

OVERVIEW OF THE BOOK

In this book, we demonstrate how the traditional approach to chemical supply leads to excessive chemical costs. We also present how Shared Savings Chemical Management programs work and how they have successfully reduced chemical volumes, reduced chemical costs, and increased chemical performance for some innovative companies. This new strategy has important implications for managers throughout a company, including in environment, health, and safety (EHS), purchasing, production, engineering, maintenance, and other departments. While Shared Savings programs are most easily applied in large operations, some or all of the components of Shared Savings have the potential to be successful in small and midsize firms.

In this chapter, we explain how managers have to cope with the *Chem-*

ical Beast that exists in most companies that use chemicals. We also demonstrate how this Chemical Beast is a product of current chemical management strategies that have not kept pace with changes in the business world. Finally, we discuss how the problems associated with current chemical management strategies arise from an outdated chemical management foundation—the traditional chemical supply relationship—and how some companies are controlling their costly Chemical Beast through an innovative chemical supply relationship known as Shared Savings Chemical Management.

The remainder of the book is divided into four parts. In part 1, we evaluate the current state of chemical management. We take an in-depth look at the Chemical Beast, the factors that contribute to it, and its impact on the company. In part 2, we present alternatives to the traditional chemical supply relationship—particularly Shared Savings Chemical Management. We illustrate how a number of companies are achieving dramatic reductions in chemical costs through this innovative approach to chemical supply. Part 3 contains five case histories of manufacturing plants that have implemented Shared Savings programs. These plants are:

1. Navistar International, Inc., Engine Plant
2. General Motors Corporation, Truck and Bus Assembly Plant
3. Ford Motor Company, Chicago Assembly Plant
4. Chrysler Corporation, Belvidere Assembly Plant
5. General Motors Corporation, Electro-Motive Division

In part 4 we discuss the implementation of Shared Savings Chemical Management, from selecting a supplier and developing a contract to monitoring success and driving continuous improvement.

THE CHEMICAL BEAST

Every company that uses chemicals has a chemical management strategy, whether by design or by default. The term *chemical management strategy* means the system used to assure the lowest total chemical costs and the best chemical performance. Unfortunately, a company's chemical management strategy typically evolves by default long after basic product and production decisions are made. As a result, it exists only as an uncoordinated, ad hoc collection of roles and responsibilities performed by a variety of personnel and departments.

Consider, for example, just a single chemical in your plant.

- How is the "best" chemical selected? Process Engineering, the equipment vendor, or even the equipment operator may select the chemical. The decision may be based on an understanding of equipment performance needs, but it is just as likely the chemical is selected because it was the "manufacturer's recommendation" or "it's what we've always used."
- How is the chemical purchased? Purchasing may search out a supplier to provide the chemical—or a similar product—at the lowest price. On the other hand, personnel may use a favorite supplier to provide the chemical. What is ultimately purchased may or may not be exactly what was requested.
- How is the chemical handled? Different personnel receive, handle, and store the chemical, perhaps employing a shake-the-drum approach to inventory control and reordering.
- How is the chemical used? Equipment operators or maintenance staff may be responsible for proper application of the chemical in the production process, but they may be guided by outdated policies and procedures—assuming that written policies and procedures exist.
- How is chemical safety assured? EHS personnel are usually called in after the fact to assure safety and compliance (monitoring, training, control, paperwork, reporting, etc.) in the use and disposal of the chemical.

Multiply this by the hundreds or thousands of chemicals used in the typical plant and you can get a Chemical Beast that is almost impossible to control and looms as a constant threat to the business.

WHY IT'S TIME TO CHANGE YOUR CHEMICAL MANAGEMENT STRATEGY

Why should chemical management strategies for the twenty-first century be based on procurement practices from the early 1900s?

The Chemical Beast is the direct result of chemical management strategies that have not kept pace with the external pressures squeezing most companies today. These pressures include:

- *Quality*—increasing customer expectations for product quality
- *Service*—increasing customer expectations for service
- *Cost*—increasing demands for efficiency and cost reductions
- *Timeliness*—rapidly changing market conditions and customer expectations
- *Regulations*—increasingly complex compliance requirements

These forces have increased dramatically in the last twenty years, yet most chemical management strategies are based on procurement practices dating from the early 1900s. Though programs such as chemical safety committees and compliance audits have been layered on top of this old procurement foundation, chemical management strategies remain rooted in a bygone business era.

As a result, many companies show signs of the Chemical Beast:

- *Controlling chemical purchasing is a nightmare*—It's hard to keep track of who has purchased what.
- *EHS compliance for chemicals is overwhelming*—Maintaining current material safety data sheets (MSDSs), employee training, and environmental reporting is a never-ending struggle.
- *Regulatory changes are hard to keep up with*—It's difficult to be certain which laws and regulations apply to which chemicals.
- *Improvements in chemical technology are hard to keep up with*—Improving chemical technology is essential, but it can be expensive, time-consuming, confusing, and risky.
- *Chemical expenses continue to increase*—The price of chemicals just keeps going up.
- *EHS expenses continue to increase*—As compliance costs rise, EHS places an increasing burden on the company's bottom line.

In an effort to resolve these escalating chemical management problems, a few companies have decided to radically change their chemical management strategy. Rather than implement more add-on programs and call in more consultants, they have fundamentally changed their approach to chemical management. These companies have begun purchasing *chemical performance* rather than chemicals from their suppliers. It is chemical performance, not chemicals, that add value to their products. This strategy requires suppliers to change their role in the relationship. The concept is simple. For suppliers to increase their own profitability, they must improve chemical performance, not chemical volume. For the companies and the chemical suppliers that have been able to adopt this new way of thinking, the financial rewards have been significant.

WHY SOME COMPANIES OVERLOOK THEIR CHEMICAL SUPPLY RELATIONSHIPS

Hidden chemical costs can be many times larger than the purchase price.

Despite the connection between chemical supply relationships and the Chemical Beast, most companies overlook their chemical supply relationship as a possible cause of their chemical problems, let alone consider their supplier relationship as a potential resource for solutions. We found several reasons for this perception. First, most managers do not know the true cost of chemicals in their company. Chemical costs are hidden in the company's accounting system. These costs include chemical acquisition, inventory, disposal, regulatory compliance, and the cost of chemical failure. In many companies, such costs can be several times greater than the chemical purchase price. Second, many managers define their core business too broadly, assuming that any activity important to production must be part of their core business, including chemical management. Third, managers often overestimate their ability to effectively manage chemicals at a competitive cost. They spend considerable time and resources developing in-house chemical expertise. Each of these problems is discussed in greater detail in chapter 3. Underestimating the cost of chemical use and including chemicals as part of their core business creates a form of managerial myopia where the need for or potential value of external expertise in controlling the Chemical Beast goes undetected.

Even when managers recognize a need for external chemical expertise, they frequently do not approach their chemical supplier or consider the chemical supply relationship as a primary source of assistance. This is because the nature of traditional supply contract creates an inherent adversarial relationship between chemical user and chemical supplier (see chapter 4). Every aspect of the traditional relationship contract, from the financial incentives to the division of responsibilities, generates a conflict of interests. Few managers would trust their chemical suppliers for extensive assistance in reducing chemical problems and their associated costs.

CHEMICAL SUPPLY ALTERNATIVES

Shared Savings aligns the interests of chemical supplier and chemical user.

Fortunately, some managers have recognized the significant impact the chemical supply relationship can have on their company. Together with a number of chemical suppliers, they have developed alternative supply strategies that perform better and return greater value than the traditional approach to chemical procurement. These alternatives cover a wide range of relationships. Many chemical suppliers today offer chemical management programs, but the benefits to the company can vary dramatically. It is useful to evaluate each of these alternatives and select the one that is best for your company. A simple way to organize the alternative chemical supply relationships is by Supplier Services and Supplier Revenue. A more detailed analysis of alternative strategies is presented in chapter 8.

Supplier Services

Supplier Services are those chemical management activities performed by the chemical supplier. At one end of the spectrum is the traditional chemical supply relationship, which may offer timely delivery of chemicals as the supplier's only service. Further along the spectrum, logistic services are often added to the supplier's responsibilities. These services typically include automated resupply, electronic data interface (EDI), and inventory management. Services related to regulatory compliance and reducing EHS problems can also be included in the supplier's responsibilities. These may include management of material safety data sheets (MSDSs), employee training, and implementation of compliance programs for selected chemicals. Finally, in more advanced chemical supply relationships, application services may be provided. These services are directed toward improving the performance of the chemical as it is applied in the production process and can range from monitoring and maintaining the quality of the chemicals on the shop floor to developing new chemicals to meet specific production requirements.

In our research, we have observed that as a chemical supplier's service responsibilities increase, the supplier's activities tend to move from *outside* the plant (providing delivery and product labeling services) to *inside* the plant (providing inventory control and process monitoring services). The supplier's role changes from responding to the chemical user's demands to proactively bringing problems and solutions to the attention of the company's personnel.

Supplier Revenue

Supplier Revenue is the fee structure whereby a supplier is paid; it plays a major role in defining the supplier's incentives. Most payment methods can be grouped into four levels:

- Level 1. **$/lb**—This traditional way chemicals are bought on a volume basis produces a transactional relationship where chemical price is the dominant factor. The supplier makes more money by increasing chemical volume.
- Level 2. **$/lb + Services**—The chemical user may pay higher prices but expects additional supplier services. The supplier makes more money by providing better services and increasing chemical volume.
- Level 3. **Management Fee**—Supplier services are separated from chemical price to reduce the incentive for suppliers to continuously increase chemical volumes. Chemical costs are passed through to the chemical user and chemicals may be obtained from other suppliers.
- Level 4. **Shared Savings**—The chemical user no longer buys chemicals; instead, the supplier is paid a fixed fee per month or per unit of product manufactured. Supplier revenue is no longer linked to chemical volume. Reducing the volume of chemical needed by the user reduces supplier costs, increasing supplier profits. In essence, chemical waste is turned into savings that are shared between chemical supplier and chemical user.

At each level of this hierarchy of revenue schemes, supplier compensation is based increasingly on the value brought to the chemical user rather than the volume of chemicals supplied. The chemical user moves from buying chemicals to buying chemical performance. This aligns the interests of the chemical supplier more closely with the interests of the chemical user to reduce chemical volumes and increase chemical performance. Suppliers have a financial incentive to decrease chemical usage because this increases their profits.

Chemical Management

Generally, movement to a higher level of the Supplier Revenue scheme is accompanied by increased Supplier Services. Chemical management programs typically involve, at a minimum, a $/lb + Services revenue arrangement with an emphasis on logistic services, including inventory management. However, more comprehensive chemical management programs, particularly Shared Savings chemical management, can produce far greater value and financial reward. Fig. 1-1 provides a simplified view of the difference between traditional chemical supply programs and chemical management supply programs.

Shared Savings Chemical Management

Shared Savings chemical management programs employ a Shared Savings revenue scheme, typically with comprehensive logistic, EHS/compliance, and chemical application services. Typical characteristics of a Shared Savings program are presented in Table 1-1.

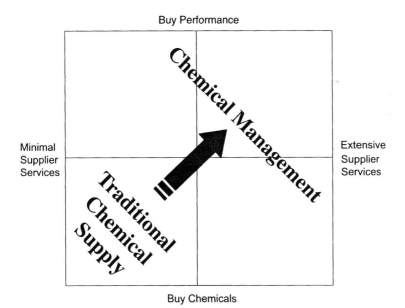

Figure 1-1. Relationship of chemical supply program to supplier services and fee structure.

Table 1-1.
Shared Savings Chemical Management Relationships—
Typical Characteristics

- User no longer buys the chemicals. They are owned by the supplier until used in the production process.
- Supplier receives a fixed fee per month or per unit of production in exchange for chemical performance.
- Supplier profits through chemical volume and cost reduction, not chemical sales.
- Supplier provides on-site chemical management, including comprehensive logistic, EHS/compliance, and chemical application services.
- One supplier serves as a primary, or Tier 1, chemical manager, overseeing the supply of chemicals from Tier 2 suppliers.

The ultimate objective of Shared Savings is to buy chemical performance rather than chemicals. After all, it is performance, not the chemicals themselves, that adds value to the chemical user's products or services. All aspects of chemical performance are relevant, from its impact on product quality to the environmental threat posed by its waste. Thus, Shared Savings uses techniques that align compensation with chemical performance and open opportunities for suppliers to increase future revenues by increasing the value they bring to the chemical user.

BENEFITS FOR CHEMICAL USERS
AND CHEMICAL SUPPLIERS

Chemical management programs, and particularly Shared Savings chemical management programs, produce valuable benefits for both chemical users and chemical suppliers. Users enjoy lower costs and better performance. The lower costs cover the full range of chemical-related expenses, including procurement, inventory, usage, handling, disposal, and regulatory compliance. The number and volume of chemicals used declines. Performance improves not only for basic services, such as delivery, testing, and maintenance, but for basic production processes as well.

For the chemical supplier, chemical management provides a valuable new competitive advantage. It improves the ability to secure new customers and the ability to increase the loyalty of existing customers. Most chemical management programs evolve into renewable, or "evergreen," programs that produce reliable revenue streams and reduce marketing and sales expenses.

IMPLICATIONS FOR THE MANAGER

Managers from EHS, purchasing, engineering, and other departments are expected to cope with the Chemical Beast that exists in many companies today. Unfortunately, most chemical management strategies are designed by default; they are fragmented and cannot keep up with the external demands placed on business. As a result, most managers spend their days "putting out chemical fires," as one manager calls it. A company's chemical supply relationship is the foundation of its chemical management strategy. These relationships are typically based on century-old procurement models that no longer meet business needs.

Changing the chemical supply relationship can change the company's chemical management strategy and improve chemical performance. Improved chemical performance produces lower overall chemical management costs as well as improved performance of EHS, which leads to better product or greater value for the ultimate consumer. In the companies we studied, managers from various departments, including purchasing, EHS, and maintenance, have played pivotal roles in initiating and implementing new chemical supply relationships. They have enjoyed not only improvements in performance and lower demand for scarce resources but also gained a higher profile in the company for proactively contributing the company's business objectives and financial success.

PART I

THE CHEMICAL BEAST

2

The Chemical Beast:
The Hidden Cost of Chemicals

> The most important insight ... is the understanding that the
> acquisition cost itself is often a very small part of the total cost.
> —D.A. RIGGS AND S.L. ROBBINS,
> *The Executive's Guide to Supply Management Strategies*

As one of his twelve labors, the Hercules of Greek mythology was sent to slay the Hydra, a beast with nine heads. As soon as one head was cut off, two more would appear and take its place. The task of slaying the Hydra is sadly similar to the herculean challenge confronted by managers whose responsibilities include chemical procurement or management. As soon as one problem is solved, two more pop up to take its place.

Consider just a few examples:

- The paint shop manager must cut costs to make budget targets. He finds a supplier for a new paint detackification chemical at 25% less than he is currently paying. When the change is made, however, it produces a major upset at the waste treatment plant. Not only does it take time and money to track down the problem, but the paint shop manager insists on keeping the new chemical.
- Every year the environment, health, and safety (EHS) department

Figure 2-1. The Chemical Beast (Image by Jack Schark)

finds over a dozen drums and cans of chemicals in various parts of the plant. No one knows who ordered the chemicals or whether the material is currently being used. Many appear to be samples provided by suppliers. The manager must take time to identify each chemical and arrange for its disposal.

- Product Engineering decides to incorporate a new material into the product to improve quality by reducing corrosion. However, this addition changes the legal status of sludge from the wastewater treatment plant, requiring that it be disposed of as a hazardous waste at a much greater cost to the company.

- The purchasing department implements procedures to assure the lowest possible price for all purchased materials. However, this creates a long lag time between the purchase request for new materials and their delivery to the plant. To avoid production shutdowns, the production department begins ordering larger quantities of chemicals. This increases inventory management costs and adds to waste disposal costs, as more chemicals become outdated while in storage.

- After an extensive market search, the engineering department introduces a new cleaner that does a better job of cleaning key production machinery. However, the cleaner is highly caustic, requiring workers to wear respirators and protective clothing, which increases maintenance costs and reduces worker productivity and morale.

- A new supplier introduces an aqueous cleaner into a parts washer. However, the cleaner is incompatible with the machining fluid used in the plant and causes spotting on the parts, significantly increasing rework.

- The budget for the boiler house includes the purchase price of chemicals used for treating boiler water. Under pressure to cut budgets, the boiler house manager finds he can save money by buying a cheaper water-conditioning chemical. However, the new chemical provides less corrosion protection for the heating pipes. Within a few years, pipe maintenance expenses increase dramatically. However, the maintenance department covers pipe maintenance, and the boiler house continues to meet its budget with the use of the new water treatment chemical.

What is the Chemical Beast? What does it look like? Why do we sometimes overlook this beast? These questions are addressed in this chapter.

THE CHEMICAL COST ICEBERG

In a corporation like this, we get a kind of tunnel vision. The purchasing guy who makes the decision about who your supplier is going to be is not the same guy that has to worry about getting rid of the waste.
—*A plant environmental manager*

The Chemical Beast can have enormous financial impact on a company, usually many times greater than the purchase price of the chemicals themselves. We first address the effect of chemicals on costs, then the effect of chemicals on performance.

Imagine chemical costs for a company as a large iceberg (Fig. 2-2). The visible portion of the iceberg—the part above the water—represents the purchase price of the chemical. However, just as the visible part of the iceberg is only a small portion of the iceberg's total size, the purchase price of the chemical is only a small portion of the chemical's total cost to the company. Chemical use creates a series of hidden costs for the company, represented by the portion of the iceberg below the water.

For many companies, the hidden costs of chemical ownership (the bottom of the iceberg) can be several times larger than the actual purchase price of the chemicals themselves. In fact, one U.S. auto company estimates that the hidden costs associated with chemicals used in its assembly facilities are five to seven times greater than the purchase price of the chemicals

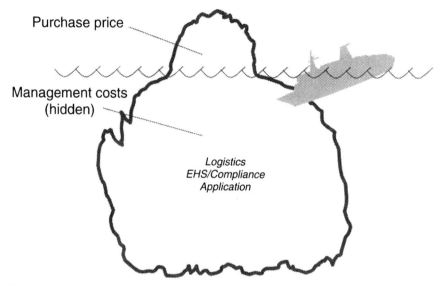

Figure 2-2. The chemical cost iceberg (adapted from Bierma and Waterstraat 1997a).

(Mishra 1997a). In other words, for every dollar this company spends on chemicals, it incurs five to seven dollars in additional chemical management costs. In general, the total cost of ownership for supplied products and materials is two to four times the purchase price for most manufacturing or assembly plants (Riggs and Robbins 1998). As with real icebergs, the portion above the water often attracts the most attention, but the portion below the water produces the greatest threat.

TYPES OF HIDDEN CHEMICAL COSTS

Almost all hidden costs can be categorized into three groups: logistic, EHS/compliance, and applications (see Table 2-1). This section presents common hidden costs.

Logistic costs

Logistic costs are the time and resources devoted to acquiring and handling chemicals. The purchasing and reorder process can be cumbersome, lengthy, bureaucratic, inaccurate, and expensive. Logistical problems can

Table 2-1.
Examples of Hidden Chemical Costs
(Bierma and Waterstraat 1997a)

Logistic	EHS/Compliance	Application
Chemical purchasing system management	Waste treatment	Maintenance costs due to chemical applications
Inventory management	Waste disposal fees	
Chemical handling	Environmental compliance	Chemical testing and quality control
	Health and Safety compliance	In-process treatment and reclamation
	Insurance	Misuse of chemicals
	Liability	
	Keeping up to date with regulations	
	Labor concerns about health and safety	

add to these costs. Chemicals may not be delivered on time or in the quantities desired. Incorrect or substandard chemicals may be sent, requiring resources to label and return. Inventory management is time-consuming and can generate waste costs from disposal of outdated inventories. Some chemicals may be lost due to poor inventory tracking or inappropriate preparation for use.

The manufacturers we studied had experienced significant logistic costs associated with use of chemicals in their production processes. In fact, the origin of Shared Savings at General Motors emerged as a strategy to reduce the routine sampling and analysis expenses (exceeding $1 million per year for some plants) incurred to monitor the quality of incoming chemicals. Other chemical managers complained about having to buy a 55-gallon drum of chemical when only a few gallons were required. Not only did the plant have to pay for the excess chemical but it also experienced an increase in hazardous waste disposal fees. The most common complaint was excessive inventory costs, which included floor space, tracking resources, and purchase and disposal costs.

EHS/Compliance Costs

EHS/compliance costs include those expenses devoted to assuring regulatory compliance as well as the overall environmental, health, and safety performance of the company. Chemicals must be packaged, labeled, shipped, and stored properly. Chemical safety data for both new and existing chemicals must be acquired, reviewed, and maintained. Employees must be trained in property safety procedures in the face of the potential chemical hazards. Employee exposures must be monitored, recorded, controlled, and reported. Discharge permit applications and fees must be submitted to ensure compliance with local and federal regulations. Chemical releases must be estimated and reported. Emergency response plans must be developed and instituted. Finally, wastes must be properly treated and disposed.

EHS/compliance costs are often difficult to fully recognize and track. For example, the Amoco Oil Company's Yorktown Refinery originally estimated plantwide environmental costs to be about 3% of total costs (excluding crude oil). However, a more thorough study found that environmental expenses actually exceeded 20% of noncrude costs. Among the many costs missed in the initial estimate, the most significant were maintenance costs associated with preventing emissions and maintaining pollution control equipment (Heller et al. 1995).

Another significant EHS/compliance expense noted by many manufacturers was data collection and reporting for the Toxic Release Inventory (TRI), as required by federal law. Dan Uhle, environmental engineer at the Ford Chicago Plant, summarized the experience of many EHS managers:

> That used to be the hardest part in doing our Form Rs—coming up with good chemical usage data. . . . Take solvents, for example. I had to look at what we bought and get records from the suppliers of all of the different materials that had VOCs [volatile organic compounds] in them. I had to make the assumption that the inventory at the beginning of the year was the same as at the end of the year. I didn't know exactly how much was scrapped and actually went out as waste paint solvent, where we had some recovery, or how much was actually emitted. So I had to make assumptions about all of that to the best of my ability, using engineering judgment.

The type of data required for regulatory reporting is often not documented and maintained by a chemical user in a usable format. Time and resources must be committed routinely by the company to ensure regulatory compliance. These costs are frequently not collected or, if collected, allocated to the overall cost of chemicals.

Application Costs

Chemical *application* costs can be a significant component of the chemical cost iceberg. These costs are incurred at the point where the chemical is used in the production process. Application costs can take a variety of forms, including the testing and maintenance needed to assure the quality of the chemicals in the process. This includes the costs of sampling, laboratory work, additives, and filtering, among others. In addition, misapplication, underuse, spills, contamination, and other associated problems are hidden in this category.

Other application costs include the cleanup of machines and work areas soiled from chemical use. For example, machining fluids are often spread as mists or splatters from screw machines, milling machines, grinding machines, and similar equipment. If not regularly cleaned, the deposited material can reduce machine performance and pose a slip hazard to workers.

Application costs for a given chemical can also include the cost of other chemicals that must be used with it. For example, the cost of solvents used to purge paint lines is a result of the specific paints used. Alternative paints may require less (or less expensive) solvents, reducing the overall cost of painting.

CHEMICAL PERFORMANCE LEVERAGE

Chemicals in a plant are like the blood in your body. If you've got bad blood, you've got problems throughout the plant. In any facility, chemical purchase costs are probably 1 or 2% of the total costs—its peanuts. But the impact of those chemicals is far greater—scrap, tool life, man-hours, downtime—the list goes on and on.
—*A chemical supplier*

Most chemical users would concur with this blood analogy. Chemicals are purchased to perform a specific function: to produce a product or a service. Poorly performing chemicals can easily disrupt production. They can also damage or shorten the productive life of equipment. They can decrease product quality, increase rework, and threaten customer satisfaction.

A useful concept, when assessing the impact of the Chemical Beast on company performance, is *chemical performance leverage*. For most chemicals, it is possible to identify a handful of ways that the chemical significantly affects quality of the process or product. For example, coolant in a typical machining operation affects performance in several key ways. First, it affects the product's finish, which influences not only the level of scrap and rework but also the value of the product. Second, it affects tool life. Poorly performing chemical products can greatly reduce tool life. Third, it affects process downtime by fouling the machine or causing dermatitis among workers. These are just a few examples.

Once you identify key performance outcomes, consider the benefits of specific improvements in these outcomes. In the coolant example, consider the benefits of a 20% reduction in scrap and rework, a 20% increase in tool life, or a 20% reduction in coolant-related process downtime. Are these improvements expected to produce only slight benefits for the company or very large benefits? (See Fig. 2-3) In some companies, a 20% improvement in tool life can produce benefits exceeding the entire coolant purchasing cost.

Of course, there is nothing special about a 20% performance improvement—any realistic value will do. The purpose of considering chemical performance leverage is to place the purchase price of the chemical in perspective relative to its value. Often, a chemical's value is confused with its purchase price. If alternative chemicals yield significantly different performance outcomes, even moderate differences in purchase price may be of trivial importance when compared to the total cost of the chemical. Conversely, decisions to save money through the purchase of less expensive chemicals can backfire if the new chemicals have worse performance outcomes.

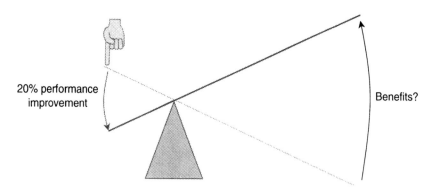

Figure 2-3. Chemical performance leverage.

THE MANY FACES OF THE CHEMICAL BEAST

One of the reasons the Chemical Beast can go unnoticed in a plant is that many of its costs are hidden. Another reason is that the Beast looks different to different people in different departments. We provide a few common examples below.

EHS

To the EHS manager, the Chemical Beast can take the form of never-ending compliance headaches. It seems impossible to keep track of *all* the chemicals used in the plant, maintain appropriate chemical safety documentation, and train workers on appropriate handling and use. New regulations seem to appear before you can get compliance programs implemented for the old ones. Annual environmental reports create a data-gathering nightmare that involves many other departments, all of which seem to have better things to do than dig up the data you need.

Perhaps it has been difficult to make headway on reducing hazardous waste generation or the discharge of air and water pollutants. Without an immediate legal threat, it is difficult to get others to see beyond today's production problems. If one department makes a change in chemicals, new problems are created for other departments. People focus on their own budgets. Even if it is costing you an arm and a leg to treat or dispose of wastes, other managers will not accept improvement ideas if it is going to raise their department's costs.

Purchasing

The purchasing department is typically charged with the responsibility of finding the lowest-priced, most reliable supplier for a specific chemical. Unfortunately, many of the decisions affecting chemical costs and suppliers are out of the hands of purchasing personnel. Production or engineering personnel often make chemical use decisions, including the brand of chemical and even the supplier. Supplier representatives visit plants, bypassing purchasing staff, and market directly to personnel on the floor. It is not uncommon for a plant to buy a dozen brands of a single chemical because individuals throughout the plant independently make such purchase decisions. The relationships between plant personnel and their favorite suppliers can be difficult to break—due, in part, to the gifts provided by many supplier reps.

In other cases, purchasing personnel may be pressured to find the lowest-cost chemical supplier in spite of other consequences. When upper management evaluates the purchasing department, a key criterion is often steady or declining purchase prices, even though lower-priced chemicals can produce other problems and costs in the plant.

Maintenance

The maintenance department is commonly faced with the dual challenge of maintaining chemical systems and using chemicals to maintain equipment, even though many of the factors that determine the success of these efforts are under the control of other departments and personnel. For example, the proper functioning of machining coolant systems relies on the production worker to operate equipment as specified, perform routine cleaning functions appropriately, and avoid contaminating the coolant with dirt, debris, trash, or other chemicals. Plant Engineering, Production, or Purchasing may specify coolants that are inherently difficult to maintain. Production schedules and budgets may not allow for proper cleaning of the equipment when coolant is changed out, resulting in rapid deterioration of the fresh coolant.

In addition to the inability to control many factors critical to success, Maintenance often does not have access to the kind of chemical expertise required. Many plants do not have a chemical engineer on staff. If there is a chemical engineer, he or she often has too many other responsibilities to spend time sorting out the causes of chemical maintenance problems.

The result is problems with equipment and chemical systems for which the maintenance department may take the heat but over which it has little control.

Production

For production departments, chemicals often mean delay and defects. Many pieces of equipment rely on chemicals, such as lubricating and hydraulic oils, in order to operate correctly. Other equipment, such as parts cleaners, uses chemicals as part of their operation. Problems occur when the wrong chemicals are used, the chemicals are not properly maintained, or when two or more chemicals are incompatible.

One result is process downtime and production delays. The breakdown of critical equipment can stop an entire production line. More time may be lost to unscheduled maintenance. Some chemicals may even cause problems, such as dermatitis, among the workers to the point of closing an entire production area. Another result of chemical problems is production defects. Poorly performing coolants can produce an improperly machined surface; poorly performing cleaners can result in soiled parts; poorly performing paints can create a pitted or blemished finish.

Wastewater Treatment

Wastewater treatment system operators must deal with the results of chemical decisions made throughout the plant. In addition, they must make their own chemical use decisions to optimize a complex chemical treatment system. Chemical wastes coming from different parts of the plant can create treatment nightmares. For example, detergents from cleaning operations can suspend oils from waste coolant, making the oils difficult to remove. The result is wastewater discharges that may exceed regulatory limits. Efforts to reduce the generation of cleaner and coolant waste may fail because they require additional expenditures by the production or maintenance departments, even though the plant as a whole would save money.

Wastewater Treatment is also faced with unpredictable and unannounced changes in the chemical composition of the wastes it receives. Someone may decide to dump a plating bath or change out a coolant system. A supplier may provide several drums of samples to an operator. Trials of a new paint may produce higher-than-usual solids from paint detackification. All of these changes can produce upsets, and potential regulatory violations, in a wastewater treatment system.

Plant Engineering

Many systems, no matter how well engineered, run only as well as the chemicals that are used in them. Sometimes the best chemicals are not chosen for the system. Other chemicals may be used because they are cheaper. Some chemicals may be used because "we've always done it that way." Still other chemicals may be purchased because they come from a favorite supplier. Even the best chemicals may not function correctly if they are not maintained or the system is not operated as specified.

In some cases, Engineering may be asked to design chemicals systems without the proper chemical expertise. For chemical engineers, much less mechanical or industrial engineers, it can be difficult to stay current with changes in chemistry, chemical technology, and regulatory requirements. Given the range of demands on Engineering, the time to research and understand new chemical developments is hard to come by.

IMPLICATIONS FOR THE MANAGER

Every chemical user has a Chemical Beast. Consider the Beast that lurks in your own company. Begin by trying to imagine the size of your chemical cost iceberg. For each chemical used in your company, consider the following questions:

- How much time, effort, and resources are devoted to purchasing, handling, and storing chemicals?
- Are chemicals often purchased on a rush basis because inventories have gotten too low?
- Do chemicals in inventory go out of date because inventories have grown too large?
- How much time, effort, and resources are devoted to regulatory compliance activities or solving EHS problems?
- Has the company received negative publicity or faced enforcement actions over EHS issues?
- Does the company use treatment or special disposal services to handle chemical wastes?
- Does use of the chemicals require expensive equipment or additional training for workers?
- Does use of the chemicals require additional maintenance or cleanup activities?

If your company is like many others, these hidden costs may run from one to seven times the purchase price of the chemicals.

Consider also the performance leverage of the chemicals used by your company.

- Do the chemicals affects product quality?
- Do the chemicals affect production efficiency (e.g., tool life, machine downtime, energy efficiency)?

The Chemical Beast looks different to different company personnel. Each person experiences different costs and frustrations in working with chemicals. What are the most important chemical issues for personnel in the following areas?

- Purchasing
- EHS
- Production
- Engineering
- Maintenance
- Wastewater Treatment

While personnel often complain about chemical problems, it should not be surprising that most companies fail to appreciate the size and scope of the Chemical Beast. This may be true in your company as well. In the next chapter, we explore the causes of the Chemical Beast and why most companies have difficulty keeping it under control.

CHAPTER
3

Causes of the Chemical Beast

What I needed were more solutions, not more problems.
—A plant maintenance supervisor

To tame the Chemical Beast, we must understand what causes it. Many factors contribute to the Chemical Beast, and these may vary from company to company. Below, we highlight the most important factors.

THE COMPLEX CHEMICAL WEB

They didn't realize that when they made that [chemical] change it would cost $100,000 in downtime and lost production and upsets and catastrophic failure.
—A plant engineer

The connections between chemicals are surprisingly complex. A change in a chemical or its application may produce outcomes that do not appear until the distant future. Problems may emerge at other points in the process with no indication of their origin. In the examples given in chapter 2, a new cleaner produced product spotting when it mixed with the machining fluid, a new paint detackification chemical fouled the wastewater treatment process, and a cheaper water-conditioning chemical led to increased boiler pipe deterioration not observed for several years after its introduction.

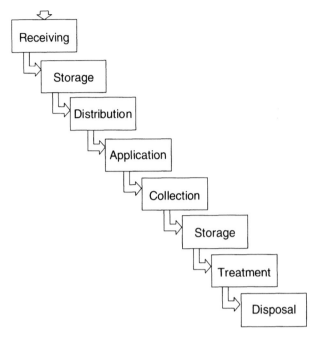

Figure 3-1. Chemical life cycle diagram (adapted from Votta et al., 1998).

Each chemical moves through its own life cycle within the plant. Fig. 3-1 presents a simple version of the life cycle of a chemical, from procurement and delivery through application and, finally, disposal. Each step in the life cycle is intimately connected with all the other steps. Decisions made at any point in the journey can have significant, and often unanticipated, effects on both downstream as well as upstream processes. In addition, solving a problem or making a change at one point can produce greater problems elsewhere in the life cycle.

CHANGING CHEMICAL TECHNOLOGY AND REGULATION

Now here, you see, it takes all the running you can do to keep in the same place. If you want to get somewhere else, you must run at least twice as fast as that.
—LEWIS CARROLL, *Through the Looking-Glass*

Chemical technology and regulation are complex and rapidly evolving. In our experience, many companies underestimate the complexity of their chemical needs and subsequently underestimate their real chemical costs. Chemicals such as coolants and cleaners often appear simple because companies have come to accept a relatively low level of performance. Optimizing performance can produce dramatic reductions in both cost and waste as well as improvements in product quality. But optimizing chemical performance requires a thorough understanding of the interaction between chemicals, equipment, and the product. Considering the hundreds of chemicals used in most manufacturing plants, this requires extensive chemical expertise. In some cases, it may even require reformulation of the chemicals used in selected processes. As a result of this lack of knowledge and expertise, most chemical systems frequently operate well below their optimum performance capabilities. A common strategy is to dump, clean, and refill chemical tanks when chemicals do not perform properly. The problem is not researched because plant personnel do not possess the time or the expertise.

The complexity of chemical regulation also continues to expand and further complicate chemical management. From shipping, storage, and use to employee health and safety to disposal and reporting, it is almost impossible for companies to be knowledgeable about all applicable regulations, much less assure that they are in compliance with them. One comment from a supervisor provides a humorous illustration of this point. He was returning to his desk after a trip to the vending machine. While reading the ingredients on the back, he remarked, "I think we need a material safety data sheet to bring this muffin into the plant!"

To paraphrase the well-known passage from *Through the Looking-Glass*, it takes all the manager's resources just to maintain the current level of chemical performance; to get ahead takes twice as much effort. A competitive edge is gained from knowing more than your competition, whether it is knowing more about your customers, knowing more about efficient production, knowing how to provide better services at lower cost, or knowing how to bring products and services to market faster. With limited resources, a company must decide where to invest its limited learning resources to generate the greatest return on investment.

The question that must be addressed is: Can a chemical user, whose core business is manufacturing a product or providing a service, justify diverting its limited resources to mastering chemical management?

FAILURE TO FOCUS ON CORE BUSINESS

We get a good waste treatment operator and then they pull him off and put him on a production process. When bells go off and all hell is breaking loose, where is he? He's at another plant putting widgets together. Same thing with the chemical engineer. He was hired to maintain and improve the paint process. But they needed him in a lab in another state. He can't focus on paint anymore. To me, that is why you do a supplier chemical management program.
—*A plant supervisor*

As international competition increases, as the pace of product innovation accelerates, and as markets and technologies expand, companies are forced to concentrate on their core business to remain competitive. *Core business* represents those products, services, or capabilities where a company possesses a competitive advantage. Focusing on core business allows a company to direct its efforts and resources toward innovation in those areas that will produce the greatest competitive strength and, ultimately, the greatest return on investment.

The same principle underlies international trade. As trade barriers are lowered and international trade expands, each country's economy shifts to those products and services for which it retains the greatest competitive advantage. In each case, the country's resources are focused on producing higher-value products and services than its competitors, resulting in increased industrial and national wealth.

The challenge is defining a company's core business. With respect to chemical use, our experience suggests that many companies define their core business too broadly. Managers often confuse *importance* with *core business*. For most manufacturers, chemicals are very important to the production process. Therefore, many managers assume that chemical management is part of the company's core business. Unfortunately, this approach to defining core business can have a significant negative impact on the company's performance.

To maintain competitiveness, a company must devote a significant proportion of its resources to *learning and innovation* in its core business areas. If chemical management is considered part of the company's core business, it competes with a company's true core business areas. Either the company's core business or the chemical management program must be compromised.

Even though many companies define chemical management as part of their core business, it is usually given a low priority compared to product design, assembly, marketing, and other activities that are truly central to the competitive strength of the company and its bottom line. Few manufacturing managers would be comfortable saying that their company is in the chemical management business. Because chemical management is not truly part of most companies' core business, there is usually a lack of support for the chemical management program. This can damage a company's competitive strength in several ways:

- *Personnel*—Companies tend to reward and promote personnel from within the true core business areas. Chemical management personnel, including environment, health, and safety (EHS) managers, rarely receive such recognition and have limited promotion and advancement opportunities. This affects the company's ability to attract and retain the best personnel.
- *Production priority*—Personnel with chemical management responsibilities are frequently diverted to other activities in order to put out fires related to core business areas such as production and product quality, leaving chemical management activities undone or poorly coordinated.
- *Innovation*—Innovation in chemical management is underfunded, as scarce learning resources are diverted to core business priorities. It becomes increasingly difficult to maintain cutting-edge status in the company's chemical systems. "The way we've always done it" is relied on as a long-term strategy.

Ironically, many managers believe chemicals are too important to exclude from a company's core business definition but not important enough to compete effectively for resources against the company's true core business functions. The Chemical Beast thrives in this type of environment.

INADEQUATE INFORMATION

We just weren't getting the right information.
A lot of times it was because our people didn't know it themselves.
—A plant engineer

Good chemical management requires a great deal of information. Operators must understand how to use chemicals properly. The connection between chemicals and their effects in the company must be clear. The usage of chemicals must be carefully monitored to identify problems. Finally, the true cost of chemical usage must be well understood. Unfortunately, these types of information are often distorted or missing entirely.

In our research, we found that the individuals responsible for using chemicals were not familiar with the chemical technology. More than once we heard environmental managers and suppliers lamenting that employees assumed that if a 5% solution was good, a 10% solution must be twice as good. This not only affected usage and waste costs but also caused problems with equipment and product quality. Lack of appropriate monitoring and cost data allowed these practices to go undetected for extended periods of time.

The relationship between chemical performance and resulting operating problems is often unclear. One reason is that effects may not be apparent until long after the chemicals are used. For example, problems with product quality may not be noted until months or years after chemicals are purchased and may never be traced to their source. In other cases, production problems may be considered an inevitable byproduct of using chemicals in the manufacturing process rather than a failure in chemical performance. For example, poor cleaning chemicals can shut down a parts cleaner and the production process that depends on cleaned parts. However, the problem may be viewed as an equipment problem rather than a failure in the chemicals themselves. The opportunity for improved equipment performance through improved chemistry may be entirely overlooked.

It may be surprising for some readers to learn that few companies can accurately document the types and amounts of chemicals required to produce one unit of product. Many managers do not even have easy access to the volume of a specific chemical used last year in their company. Unfortunately, a large number of managers are unsure of exactly which chemicals are used in their plants and for what purpose. Given the complexity of today's chemical technology and the fact that chemicals are not truly part of most manufacturers' core business, the lack of such information should not be surprising. It is time-consuming and expensive to develop, implement, and maintain accurate chemical tracking systems.

Cost accounting systems also contribute to the problem. Many chemical-related costs are pooled and recorded as overhead, particularly those costs related to EHS, such as waste treatment, disposal, and documentation. These costs may be allocated to products or divisions through the accounting system, but such allocations typically involve cost formulas based on direct labor hours or square feet of floor space. These accounting practices do not relate the use of the chemicals to the costs they generate.

For example, consider a plant operating a wastewater treatment facility at a cost of one million dollars per year. One production division spends $200,000 per year on a chemical that produces 50% of the load on the wastewater treatment plant. Divisions are allocated to wastewater treatment overhead on the basis of square footage. Because the division occupies 10% of the plant floor space, it is allocated $100,000 of the water treatment facility costs, even though it creates $500,000 in wastewater treatment costs. To save money, the division management presses the purchasing department to find a cheaper chemical. Purchasing finds a supplier that has a 20% lower chemical purchase price, but the chemical will increase the division's load on the water treatment plan by 20%. The division purchases the cheaper chemical; their chemical purchase price declines $40,000 while their overhead allocation increases only $20,000 for a net operating savings of $20,000. The plant as a whole, however, loses. It saves $40,000 on chemical purchases but loses $100,000 in increased treatment costs for a net loss of $60,000.

CHEMICALS CUT ACROSS DEPARTMENTS

I had a lot of personal frustration with trying to get a hold on processes and control of the waste treatment systems. With all these independent groups that were doing their own thing, it was basically chaos. People are so focused on what it takes to make the product—particularly their product or component—they don't see the big picture.

—A plant engineer

Chemical responsibilities are typically scattered across an array of departments. The following scenario is not uncommon:

- Engineering specifies the type of chemical required in a process.
- Purchasing selects the vendor and specific chemical to be delivered.
- Maintenance is billed for the chemical.
- Production personnel actually use the chemical.

- Engineering designs the wastewater treatment process to treat the chemical when it becomes waste.
- Maintenance actually operates the wastewater treatment plant.
- EHS is responsible for regulatory compliance at every step in the process.

Companies today are realizing many of their problems are caused by a lack of cross-functional coordination. Traditional organizational structures are designed to focus departments on a narrow set of responsibilities in the hope that each department will develop a high level of expertise in managing those responsibilities. Department budgets are intended to focus activities more narrowly and promote cost control within each department. The downside is that these organizational structures and policies inhibit interdepartmental or cross-functional cooperation. The resulting behavior is individual departments pursuing their own performance and production objectives, sometimes to the detriment of the organization as a whole.

This lack of cooperation provides more food for the Chemical Beast. The complex relationship between chemicals and the production process requires careful coordination of chemical management responsibilities and functions, yet the common organizational structures and policies cannot provide this—and, in some cases, work against it. As shown above, one department's decisions to reduce its chemical operating costs or improve process performance can significantly increase costs and reduce performance elsewhere in the company.

OUTDATED CHEMICAL
SUPPLY RELATIONSHIPS

It's a moving target—a shell game. One supplier would cut cost with a new chemical, but then we had an upset with the waste treatment system. Now I got a problem over there and it's gonna take more chemicals to treat it. Then the pipes start gumming up and valves stick. Chemical costs just keep going up. The concept is that you make one supplier responsible—then there is none of this shell game.
—*A plant supervisor*

Chemical purchase practices can have a profound effect on the Chemical Beast that harasses a company. Certain purchasing strategies not only feed the Chemical Beast but also compound other contributing factors. The traditional approach to buying chemicals and dividing responsibilities between chemical user and chemical supplier can promote waste and inhibit innovation. The role of the chemical supply relationship can be so critical to a company's chemical management success that we devote chapter 4 to exploring the traditional approach to chemical supply and how it is an inherently wasteful relationship.

IMPLICATIONS FOR THE MANAGER

Given the number of factors contributing to the Chemical Beast, it is not surprising that most companies have great difficulty keeping it under control. To determine the causes of the Beast in your own company, consider the following questions:

- Do chemicals affect each other or the production process in complex ways; is the interaction poorly understood?
- Is the company easily able to keep up with new chemical technologies and regulations without devoting significant time and resources to the effort?
- Does company management consider chemicals to be within their core business? For example:
  ⇒ Are personnel involved in chemical management recognized and promoted to the same extent as personnel involved in production, product design, or marketing?
  ⇒ Does chemical management have the same priority as production?
  ⇒ Does innovation in chemical technology occur as frequently as innovation in production or product design?
- Do company personnel have a good understanding of:
  ⇒ the number of chemicals used?
  ⇒ how chemicals are used?
  ⇒ the volume of chemicals used, by machine?
  ⇒ the types and volumes of wastes produced?
  ⇒ the costs of these chemicals and wastes?
- Do chemical responsibilities cut across departments, and do these departments have difficulty coordinating activities in order to reduce overall chemical costs?

- Are chemical suppliers actively working to minimize the company's chemical usage and chemical costs?

If your answer to the last question is no, your company may be missing one of its greatest opportunities to tame the Chemical Beast. The traditional chemical supply relationship compounds the chemical problems in a company. This is the subject of our next chapter.

4

Inherently
Wasteful Relationships

The usual first reaction is simple leverage to the maximum extent possible to squeeze every dime—to show no mercy....Not only does this fail to produce a low total cost of ownership, but [it] misses significant strategic potential.
—D.A. RIGGS AND S.L. ROBBINS,
The Executive's Guide to Supply Management Strategies

The Chemical Beast is typically the byproduct of factors introduced in chapter 3. In our research, we found that no single factor is more important than the company's chemical supply relationship. The characteristics and incentives of the traditional supply relationship make significant improvements in chemical performance difficult to achieve. In many instances, the typical supply relationship contributes to continuous *increases* in chemical use and cost.

The traditional chemical supply relationship contributes to the growth of the Chemical Beast through a series of steps, beginning with the principle of *adversarial interests* (see Fig. 4-1). These steps form a cycle that is self-reinforcing. They lock the chemical user into an underperforming (and, in some cases, a nonperforming) supply relationship while feeding the ever-growing Chemical Beast.

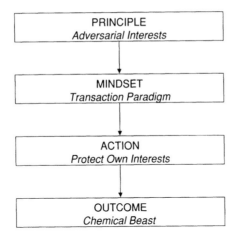

Figure 4-1. Traditional chemical supply relationships.

THE PRINCIPLE: ADVERSARIAL INTERESTS

*In one basic respect, the supplier's objectives are always at odds
with the client's: one's revenue is the other's cost.*
—*V. Kapoor and A. Gupta*

Traditional chemical supply relationships are primarily sales relationships based on the underlying principle that chemical suppliers and chemical users have inherently adversarial interests. In the language of business strategists, it is a zero-sum game—for one side to gain, the other side must lose.

This adversarial principle is reflected in (and reinforced by) the financial terms of the supply relationship. Chemicals are typically purchased on a dollars-per-pound or dollars-per-gallon basis. Under this arrangement, the supplier increases profit by increasing the volume of chemicals sold (see Fig. 4-2). The financial incentives continuously drive the supplier to increase chemical sales in order to increase profit.

This does not mean that suppliers consciously promotes wasteful practices to their customers (though it happens). However, when a company buys chemicals on a dollars-per-volume basis, it establishes fundamental incentives for the supplier. For chemical salespersons operating under a traditional supply relationship, the primary incentive in their professional life is to increase the sales volume of chemicals. Some suppliers actually give a Golden Drum Award to sales staff who sell the greatest volume of chemical each year. New products, technologies, or practices that improve chemical

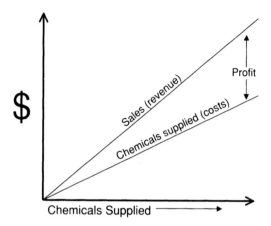

Figure 4-2. Traditional supplier relationship—a supplier's incentive to increase chemical volume (Bierma and Waterstraat, 1997a).

efficiency or reduce cost are viewed as a potential threat to sales, though chemical suppliers may occasionally support these volume-reduction or cost-saving strategies if they perceive that another supplier is competing for their business.

This behavior is not the result of a "bad" supplier. Instead, *it is a natural outcome of a supply relationship that links supplier revenue directly to chemical volume.* It produces what we call the *volume conflict* between chemical user and chemical supplier. The supplier's financial interests are in direct conflict with the chemical users' desire to reduce chemical volumes.

THE MINDSET: TRANSACTION PARADIGM

The principle that chemical supplier and chemical user have inherently adversarial interests leads to what we call a *transaction paradigm*—a belief that the primary value of the relationship is in the transaction: the exchange of chemicals for money. The transaction paradigm is built on a number of practices that feed the Chemical Beast: a focus on purchase price, cost shifting, a loading-dock division of chemical responsibilities, and a short-term time horizon.

Focus on Purchase Price

Meeting budgets by squeezing the top of the iceberg can simply shift
equal or greater costs to the bottom of the iceberg. Companies
succeed by eliminating costs, not shifting them around.

In a traditional supply relationship, value is perceived to arise in the sales transaction—the exchange of chemicals for money. Given this, it is not surprising that many buyers develop an obsession with purchase price—the lower the purchase price, the greater the value of the transaction. There are often powerful incentives within a company to create and maintain this focus on purchase price. Many departments attempting to meet or reduce budgets seek suppliers offering lower prices. Just as chemical sales personnel are given the Golden Drum Award for increasing sales, company purchasing agents are often recognized for obtaining the lowest price on the goods and services they purchase for the company. Unfortunately, purchase price is easy to measure; quality and total cost are not.

Cost Shifting

Attempts to reduce chemical costs can result in a problem we call *cost shifting*. Saving a few dollars on the price of chemicals can simply shift equal or greater costs to the bottom of the iceberg as logistic, EHS/compliance, and chemical applications costs. Suppliers may seek to make up lost revenue by cutting services or quality. Some companies shift chemical management costs, such as inventory control or waste reclamation, without corresponding measures to improve efficiency in these activities. Shifting costs from buyer to supplier results in the same (or greater) costs for the entire supply chain to the customer. Companies succeed in a competitive marketplace by *eliminating* costs, not just shifting them around.

Loading-Dock Division of Chemical Responsibilities

Why should the chemical user take primary responsibility for chemical performance?

Traditionally, chemical responsibility and ownership changes hands on the loading dock of the manufacturer as a chemical drum is unloaded from the supplier's truck and received by the manufacturer's employee. This is a convenient point for transferring a chemical's physical and legal responsibility while simplifying accounting practices, but the loading-dock mentality often produces enormous inefficiencies and hidden costs for the chemical user. This division of responsibilities does not take advantage of each company's core competencies; it also distances the supplier from ultimate responsibility for chemical performance.

For most companies, chemical management is outside their core expertise, and the time and resources they devote to chemical management represent a drain on their core business resources. In essence, it is an attempt by the chemical user to duplicate the resources and expertise of their chemical supplier—whose core business *is* chemicals and chemical management.

Contrary to common perception, responsibility for chemical performance falls largely on the manufacturer rather than the chemical supplier in a traditional supply relationship. Once suppliers are paid for a chemical delivery, they are financially removed from ultimate responsibility for the performance of their chemicals. Though a supplier may be called in to address chemical performance problems, it is generally after the fact and only indirectly related to the supply transaction. This approach to a chemical supply relationship encourages suppliers to focus more on chemical sales than performance.

Many companies believe that they hold their suppliers responsible for the performance of purchased chemicals, but how do they accomplish this? Typically, the manufacturer must closely monitor chemical performance, identify problems, and then seek corrective action from the supplier. To obtain this corrective action, the manufacturer must use an expressed or implied threat to switch suppliers (which creates other problems). Because payment occurred at the time of chemical delivery, no direct financial impact threatens the supplier as a result of poor chemical performance other than the potential loss of future sales. The manufacturer must first determine that the problem was chemical-related and trace it back to the appropriate vendor. Current chemical tracking systems make this difficult and time-consuming. Even when a supplier is called in, it may be several days before a sales representative visits the company and several more days before the problem is resolved. This can result in a significant loss of production time. In traditional supply relationships, most of the chemical performance burden and responsibility falls squarely on the shoulders of the chemical user.

Short-Term Perspective

Given the emphasis on the chemical sales transaction and, particularly, the importance of purchase price, it is not surprising that supply relationships are perceived as temporary arrangements. Chemical users can easily switch suppliers to take advantage of lower prices, using this option as a means of controlling chemical prices and threatening their suppliers into action. This creates a supply relationship with a short-term focus. From the supplier's perspective, committing equipment, financial resources, or personnel to long-term projects is too risky. This type of relationship also inhibits the flow of useful information because the supplier has no assurance that the customer will not simply use one supplier's ideas while purchasing chemicals from another supplier at a lower price. What many chemical users fail to recognize is that most costs in the chemical iceberg can be reduced only through long-term efforts. Chemical user and chemical supplier must make mutual, long-term investments and share their expertise.

ACTION: PROTECT YOUR OWN INTERESTS

The sales transaction paradigm, driven by a dollars-per-pound financial relationship, leads to a wide variety of activities in which both chemical user and chemical supplier attempt to further their own financial interests, often at the expense of one another. Below are a few of the many examples related to us by manufacturers and chemical suppliers alike.

The Ever-growing Inventory—A manufacturing plant used a local supplier to provide its maintenance paints. Whenever a maintenance paint project was proposed, the supplier found a multitude of reasons to justify why the volume of paint needed should be greater than the volume estimated by plant personnel. The supplier also found reasons why the leftover paint from previous jobs was not usable. The plant routinely had a large volume of paint left at the end of the job, which could not be returned because it was a specially mixed color. Over time, the plant developed an enormous inventory of leftover maintenance paints that had to be periodically disposed of as a hazardous waste. [The rest of the story is: Once the company switched to a Shared Savings supply strategy, which eliminated the transactional incentives, not only did the paint supplier provide the correct amount of paint to the nearest gallon, they also found many ways to effectively use the inventory of previously leftover paint.]

Dump and Fill—One company used a large volume of coolants and oils, from machining fluids to heat-treat quenches and hydraulic oils. They utilized a number of suppliers, who competed with each other for a greater share of the plant business, to control costs. No one supplier had enough volume to justify significant technical support to the plant. Although all the suppliers had a keen interest in increasing sales volume, they provided only minimal investment of time and resources. It became a standing joke with the plant personnel that the only advice you could get from a supplier was "dump and fill." If the plant was having any sort of problem with a fluid, the supplier explained that the plant must have contaminated it. The best solution was to dump the old fluid and fill it with new. [The rest of the story is that three to four tanker trucks used to arrive on a regular basis to remove waste chemicals. Now, under Shared Savings, only one tanker truck is required. The supplier's Shared Savings chemical manager said his only regret was that the savings generated from the reduction in disposal services was not included in his performance bonus.]

Never Say Good-bye—A manufacturing plant decided to switch chemical suppliers and notified its current supplier of the date its contract would be terminated. On the last day of the contract, the supplier arrived at the plant with a semi-trailer full of chemicals that had not been ordered and proceeded to unload it into the plant's warehouse. Several days later, the plant received a bill for the entire load. After a flurry of threats to the supplier, the manufacturer not only paid for the chemicals but paid to have them removed. The cost of potential litigation would have exceeded the cost of the chemicals and their disposal. [The rest of the story is that once the manufacturer switched to a Shared Savings supply strategy, the new supplier helped the manufacturer inventory leftover chemicals. The new supplier used the chemicals that were still applicable in the plant before switching to their own chemicals. For those chemicals that the manufacturer no longer used, the supplier found other companies to purchase them.]

Quality Taboo—In a plant where products were painted a variety of colors, paint guns and hoses had to be flushed between the frequent color changes. Despite significant cost and atmospheric emissions from the flush solvent, the supplier insisted that each flush required a minimum of 10 gallons of solvent. Anything less would risk spotting on the painted surface, requiring costly rework or scrap product. Rather than risk its quality reputation, the plant continued to use 10 gallons of solvent for every gun on every flush. [The rest of the story is: After the plant adopted a Shared Savings chemical supply program, the supplier "discovered" that it could blow a plug through the line, squeegeeing out most of the paint, and then purge the line with only 2 gallons of solvent—an 80% reduction in chemical usage and waste.]

THE CHEMICAL BEAST GROWTH CYCLE

As we have seen, the traditional chemical supply relationship begins with the principle of adversarial interests, which leads to the transaction paradigm, which produces practices to protect one's own interests. This behavior ultimately feeds the Chemical Beast. To make matters worse, the effects of the Chemical Beast reinforce the principle of adversarial interests, starting the cycle all over again (see Fig. 4-3). From the chemical user's perspective, the supplier's practices support the perception that they cannot be trusted. The common reaction is to attempt greater control of the supplier by increasing the use of threat. The threat of switching suppliers is seen as the ultimate (and possibly the only) control resource available to influence the suppliers.

Unfortunately, the manufacturer's use of threat only reinforces the supplier's perception that the manufacturer cannot be trusted. So, the

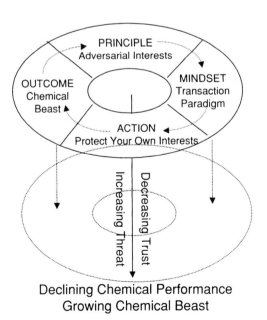

Figure 4-3. The Chemical Beast growth cycle.

chemical user's attempts to continuously reduce purchase price are met with hidden supplier strategies to maintain profits. It further reduces the flow of useful information and the willingness to mutually invest in long-term improvements. All of these actions contribute to the growth of the Chemical Beast.

The use of threat to change or control supplier behavior is effective only as long as the threat is sustained. Ironically, it forces the supplier to focus resources on reducing the threat of being replaced rather than on improving processes or reducing costs. The supplier may improve performance, but only up to the point that reduces the implied threat. There are no financial or other benefits for the supplier to pursue further performance improvements. To motivate the supplier to continuously improve performance, the manufacturer must continually use additional threats. As noted previously, this puts the actual responsibility for performance on the manufacturer rather than the chemical supplier.

IMPLICATIONS FOR THE MANAGER

To what extent does your company's chemical supply relationship contribute to your Chemical Beast? Could a change in the way your company buys chemicals help tame your Chemical Beast? Consider the following questions:

- Is your supplier's revenue linked to the volume of chemicals you use? (Are you buying chemicals by the pound or gallon?)
- Does the purchasing decision tend to focus on finding the supplier with the lowest purchase price?
- Does your company attempt to shift costs to the supplier (such as chemical testing or inventory management) instead of reducing or eliminating such costs?
- Do chemical management activities tend to switch from supplier personnel to your company's personnel at the loading dock?
- Is your company hesitant to make long-term investments with your chemical supplier to reduce costs or improve performance?
- Have your chemical suppliers acted in ways that benefit themselves at your company's expense?
- Do chemical problems seem to be locked in a never-ending cycle, with the company blaming a supplier and one supplier blaming another?

If any of these problems sound familiar, your company can benefit from improvements in its chemical supply strategy. Fortunately, a number of superior alternatives to the traditional supply relationship are available. However, before discussing alternatives, it is useful to clarify the role of chemicals in a company's competitive success. How much damage does the Chemical Beast do in your company? This is the issue we address in the next chapter.

CHAPTER

5

Total Cost of
Chemical Ownership

In our business, chemicals seemed like a very small part of our costs.
But we were wrong. And it was out of control.
—*A plant maintenance supervisor*

Many companies underestimate the impact of their chemical supply strategy because they underestimate the impact of their chemicals. Purchase price is usually a poor measure of chemical impact. Chapter 2 showed how the chemical cost iceberg is a useful way to visualize the impact of a chemical on overall costs. However, it provides little insight to the sources of those costs or how they affect the cost of specific production operations. In this chapter we introduce the *chemical life cycle* and *total cost of chemical ownership* (TCO) as useful tools for clearly seeing the Chemical Beast. In the next chapter we show how TCO affects product costs. Then, in chapter 7, we link chemical costs to business value and ultimate business success. Once the Beast is clearly defined, we can begin to tame it.

THE CHEMICAL LIFE CYCLE

The chemical cost iceberg represents the total cost of ownership for a chemical. TCO is a concept familiar to most purchasing managers, but it is

48

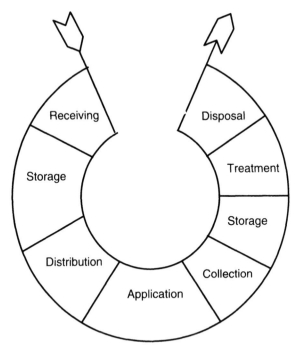

Figure 5-1. In-plant chemical life cycle (adapted from Loren 1996, and Votta et al. 1998).

more commonly applied to materials and equipment, such as computer technology. However, the TCO concept is applicable to chemical use and can be informative in understanding the financial impact of chemicals.

In Chapter 3, we briefly introduced the chemical life cycle. A chemical enters the plant, flows through a series of operations, and leaves the plant, either as part of the product or as waste. A common life cycle for chemicals can be pictured as in Fig. 5-1. The chemical is received and stored in inventory. It is removed from inventory and transported to its point of use. The chemical is applied to the production process. For coolants, this might be usage in a screw machine. For a boiler water treatment chemical, this might be mixing with the boiler feed water. For an aqueous cleaner, this might be usage in a parts washer. While in use, the chemical may be monitored and adjusted. In some cases, in-process reclamation, such as filtering or separating, may be applied. Once the chemical is removed from the process, it may be stored and treated, and it is then disposed of.

COSTS AT EACH STEP IN THE LIFE CYCLE

With a few exceptions, such as paints and plastic resins, most chemicals do not become part of the product. Instead, once they are no longer usable in the process, they must be collected, stored, treated (if necessary), and, finally, disposed of. Disposal may take the form of emissions, discharges, waste haulage, or any other means by which the chemical leaves the plant.

Each step in the chemical life cycle incurs costs (Fig. 5-2)–labor for receiving and transporting chemicals, equipment for storing and pumping chemicals, the occupation of building space, treatment costs, disposal fees, and many other expenses. Table 5-1 provides typical examples of costs incurred at each step in the chemical life cycle.

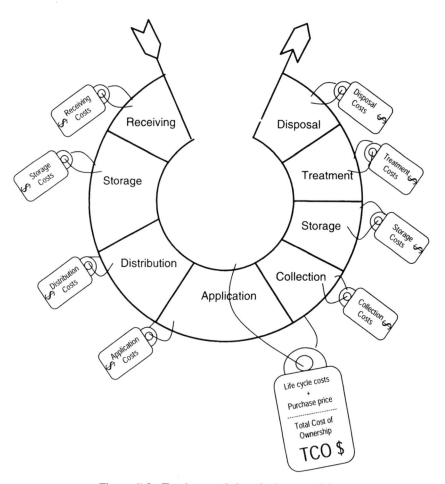

Figure 5-2. Total cost of chemical ownership.

Table 5-1.
Examples of Costs at Each Step in the Chemical Life Cycle

Receiving (includes all activities leading up to and including the receipt of chemicals and the payment of invoices)

Material Costs—value of spilled material

Logistic Costs
- Finding vendors
- Negotiating price
- Ordering
- Unloading
- Quality monitoring and laboratory analyses
- Checking and paying invoice

EHS/Compliance Costs
- Chemical screening and clearance
- Assuring transportation compliance
- Container labeling
- Spill containment construction
- Spill containment activities
- Collection and maintenance of material safety data sheets
- Worker safety training and equipment
- Emergency response planning
- Emergency response equipment
- Annual reporting

Storage

Material Costs—value of spilled material

Logistic Costs
- Drum and tote handling
- Pumping and transfer equipment
- Storage equipment
- Electricity for pumps
- Building occupancy costs

EHS/Compliance Costs
- Container labeling
- Spill containment contraction
- Spill containment activities
- Worker safety training and equipment
- Emergency response planning
- Emergency response equipment
- Emissions control from transfer and storage equipment

Distribution

Material Costs—value of spilled material

Logistic Costs
- Drum and tote handling
- Pumping and transfer equipment
- Electricity for pumps

Table 5-1. (*Continued*)

EHS/Compliance Costs
- Container labeling
- Spill containment construction
- Spill containment activities
- Worker safety training and equipment
- Emergency response planning
- Emergency response equipment
- Emissions control from transfer equipment

Application

Material Costs—value of spilled material
Application Costs
- Worker training in proper application
- Equipment for in-process storage and application of chemical
- Electricity for equipment
- Quality monitoring and laboratory analyses
- In-process reclamation or treatment

EHS/Compliance Costs
- Container labeling
- Spill containment construction
- Spill containment activities
- Worker safety training and equipment
- Emergency response planning
- Emergency response equipment
- Emissions from application equipment

Collection

Material Costs
- Value of spilled material
- Value of material in product

Logistic Costs
- Drum and tote handling
- Pumping and transfer equipment
- Electricity for pumps

EHS/Compliance Costs
- Container labeling
- Spill containment construction
- Spill containment activities
- Worker safety training and equipment
- Emergency response planning
- Emergency response equipment
- Emissions control from transfer equipment

Storage

Material Costs—value of spilled material
Logistic Costs
- Drum and tote handling
- Pumping and transfer equipment

Table 5-1. (*Continued*)

- Storage equipment
- Electricity for pumps
- Building occupancy costs

EHS/Compliance Costs
- Container labeling
- Spill containment construction
- Spill containment activities
- Worker safety training and equipment
- Emergency response planning
- Emergency response equipment
- Emissions control from transfer and storage equipment

Treatment

Material Costs—value of spilled material

EHS/Compliance Costs
- Treatment labor
- Treatment equipment
- Utilities
- Chemicals used in treatment
- Building occupancy costs
- Container labeling
- Spill containment construction
- Spill containment activities
- Worker safety training and equipment
- Emergency response planning
- Emergency response equipment

Disposal

Material Costs
- Value of spilled material
- Value of material in waste

EHS/Compliance Costs
- Finding waste handling contractor
- Negotiating price
- Assuring compliance of contractor facilities
- Disposal fees
- Permit fees
- Discharge fees
- Container labeling
- Spill containment construction
- Spill containment activities
- Worker safety training and equipment
- Emergency response planning
- Emergency response equipment
- Obtaining permits
- Completing paperwork and maintaining record

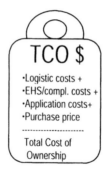

Figure 5-3. The real chemical price tag.

THE REAL CHEMICAL PRICE TAG

The real chemical price tag is not just the material cost (purchase price) but also all of the logistic, environment, health, and safety (EHS)/compliance, and application costs incurred at each step in the chemical life cycle (Fig. 5-3). Though current accounting practices do not reveal this chemical price tag, it is the true cost incurred by the company every time a gallon of coolant is added to a milling machine or every time a pound of caustic soda is used in a water treatment process. To understand the Chemical Beast, plant personnel must see the real chemical price tag and how their activities contribute to that cost.

IMPLICATIONS FOR THE MANAGER

What is the TCO for chemicals in your company? Consider these questions:

- What does your company spend each year on chemicals?
- What does it take to purchase a chemical (approvals, paper work, bidding, etc.)?
- Do you get the right chemical for the job, or is it more often a case of trial and error?
- Are chemical inventories often too high or too low?
- What proportion of your company's EHS budget is related to the chemicals you buy?
- What does your company spend on waste treatment and disposal?
- If your chemicals performed significantly better, how much would that benefit the company?

- How often do you have to stop operations to replace ineffective chemicals?

If your company is similar to other plants in manufacturing and assembly, your hidden chemical costs are anywhere from one to seven times your purchase costs—or more! If the managers in your company knew the TCO, would chemical management become a higher priority?

CHAPTER

6

Chemicals Affect Product Costs

Process improvement and process optimization are the keys.
—Dr. P.N. Mishra, *Chemicals Management Program Coordinator,*
General Motors Corporation

Chemicals play an important role in determining the cost of a company's products (and services). A clear understanding of how chemicals affect product costs is necessary to identify the full reach of the Chemical Beast in limiting the competitive strength of a company. It is also needed to identify the best ways of taming the Beast.

As we saw in the previous chapter, the cost of a chemical includes not only the purchase price but all of the hidden costs that make up the total cost of ownership (TCO). These chemical costs add to product costs through the *operations* in which they are used. However, we show in this chapter that a chemical's *performance leverage* can also affect product cost by changing the *efficiency* of operations. Finally, we illustrate how *chemical expertise,* from both inside and outside of the plant, can drive down product costs.

PRODUCTION OPERATIONS— WHERE VALUE IS CREATED

Any business can be viewed as a collection of operations. Together, these operations create a product or service that has value for the customer. A single

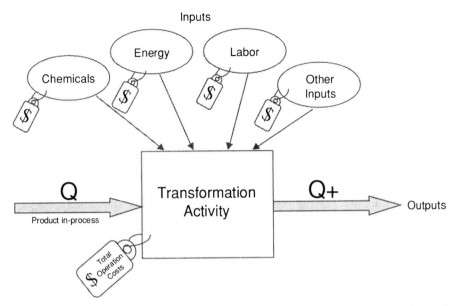

Figure 6-1. A basic production operation with inputs, transformation activity, and output.

operation involves (1) inputs, (2) a transformation activity, and (3) outputs (see Fig. 6-1). A group of related operations constitutes a *process*, and a plant may utilize a number of processes to produce a final product.

Inputs to an operation include not only the product in-process, but also component parts, materials, chemicals, energy, water, labor, etc. The transformation activity in an operation uses these inputs to increase the value of the product by enhancing its quality. In Fig. 6-1, this is shown by increasing product quality from a level Q to a level $Q+$.

The cost of a product increases as it passes through each operation. Inputs applied during the operation contribute their costs to the overall cost of the product. The total cost contribution from all inputs determines the total operation cost.

HOW CHEMICALS CONTRIBUTE TO OPERATION COSTS

Through the TCO

Each input used in the transformation activity has a cost. For chemicals, this cost, beyond the chemical purchase price, includes the costs of ordering, handling, storing, regulatory compliance, disposal, and all other chemical-related activities (see chapter 5). This total cost is referred to as the chemical's TCO. As a chemical is used in a production operation, it contributes its TCO, not just its purchase price, to the cost of the operation and, in turn, the cost of the product.

For example, consider an operation that machines a groove into a metal bar. A blank bar (Q) enters the milling machine as rough stock and exits the operation with a groove (Q+)—a higher level of product quality. Coolant is used in the machining process. Other resources are used in the operation as well, including the milling machine, milling tools, labor, and energy. The total cost of coolant ownership, as well as the total cost of ownership for the other resources, must be calculated and totaled to produce the total cost of the machining operation.

Fig. 6-2 illustrates how chemical TCO contributes to the cost of a production operation. The chemical life cycle interfaces with the operation at the application stage. Recall that each step in the chemical life cycle generates logistic and environment, health, and safety (EHS)/compliance costs. The application step generates application costs. Together with purchase price, these costs create the total cost of chemical ownership. This total cost is contributed to the operation, and the product, as the chemical is applied.

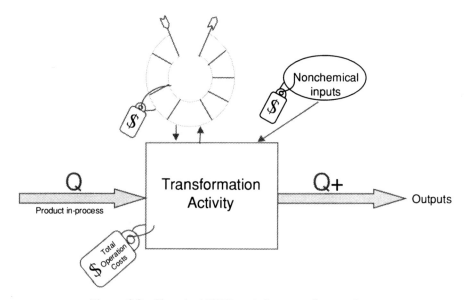

Figure 6-2. Chemical TCO and the cost of operations.

Increases or decreases in the chemical-related costs generated at any step in the chemical life cycle affect the total cost of chemicals and, subsequently, the total cost of the production operation. In our machining example, delivering coolant in totes instead of drums can reduce ordering, purchasing, receiving, and storage costs. This reduces the total cost of coolant and, ultimately, the total cost of the machining operation.

The cost of chemicals contributes to the cost of operations in another important way. Many nonchemical inputs to a production operation are produced or treated in-plant with the use of chemicals. Examples include steam, deionized water, and conditioned air. Even the cleaning and preparation of material inputs, such as sheet metal, may require the use of chemicals.

Consider, for example, the production of steam. Chemicals are used to treat boiler feed water in order to protect boiler pipes and minimize deposits. Chemicals are added to steam distribution lines for the same purposes. Even the periodic maintenance of boiler and steam lines can use chemicals for cleaning and treating pipe surfaces. The TCO for these chemicals adds to the cost of steam. As steam is used in a production operation, it contributes not only the cost of water, fuel, and equipment but also the TCO cost of steam-related chemicals. Thus, changes in the TCO cost of steam-related chemicals change the cost of steam. This, in turn, changes the cost of production operations that use steam.

In most plants, both the cost of chemicals used directly in production operations and the cost of chemicals used in producing nonchemical inputs are important in determining the overall cost of operations. Neither should be overlooked in identifying the reach of the Chemical Beast.

Through Performance Leverage

As we have shown, chemicals can affect operation costs by directly contributing their TCO to an operation or by indirectly contributing their TCO through an input that was produced with chemicals. Another important way that chemicals can affect operation costs is through *performance leverage*. As discussed in chapter 2, performance leverage is the effect that a chemical can have on the performance of an operation. One important aspect of operation performance is *efficiency*.

Operation efficiency is measured by the amount of inputs required by an operation to produce a unit of output. The greater the amount of inputs required per unit of output, the lower the operation efficiency. In many operations, chemicals can influence this efficiency. In our machining example, a superior coolant can extend tool life, reducing the total cost of tooling required to produce a quality product. A superior coolant can also reduce machine downtime, increasing the efficiency of the operation by reducing the amount of labor and equipment applied per unit of output.

As discussed in chapter 2, performance leverage can be quantified as a measure of the benefit obtained from a given level of performance improvement. If tool life, for example, was extended 20% by using a superior coolant, the performance leverage is the corresponding savings in tooling costs. If the benefits from chemical-related performance improvements exceed any increase in chemical costs, the result is lower total operation costs.

Similarly, chemicals can affect the efficiency of ancillary operations that produce nonchemical inputs. Returning to our example of steam production, chemicals significantly affect boiler efficiency by limiting the development of sediment and deposits inside boiler pipes. Decreasing sediments and deposits increases heat transfer efficiency, reducing the amount of fuel required to produce steam. The savings in fuel translate into lower steam costs. This, in turn, reduces the cost of all the other operations in the plant that require steam.

The overall impact of chemicals on operation costs is summarized in Fig. 6-3, which illustrates the four ways that chemicals can affect operation costs. First, the TCO of chemical inputs contribute directly to operation costs. Second, the TCO of chemicals used in producing nonchemical inputs (such as steam) contributes their costs through the overall cost of those inputs. Third, chemicals can directly affect the efficiency of the operation. Finally, chemicals can affect the efficiency to ancillary operations that produce nonchemical inputs.

THE PRODUCTION PROCESS— A COLLECTION OF OPERATIONS

Up to this point, we have illustrated how chemicals can affect operation costs. To understand how chemicals can significantly affect the cost of an entire production process, it is necessary to link a series of operations together. We return to the bar stock machining example. Fig. 6-4 presents the entire bar stock production process, from the receiving of raw bar stock to polished grooved bar.

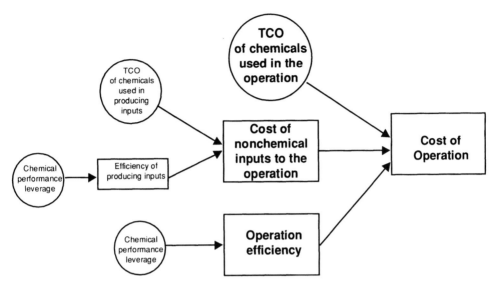

Figure 6-3. Chemical effects on the total cost of an operation (chemical effects are indicated in circles).

Chemicals are used in at least four of the operations in this process: cutting fluid is used in the bar-cutting operation, coolant is used in the machining operation, degreaser is used in the degreasing operation, and rust preventative (RP) is used in the rust prevention operation. These chemicals contribute their respective TCOs (not just their purchase price) to the cost of each step of the overall process.

Chemicals also affect the efficiency of each operation in which they are used. For example, the coolant used in the machining operation affects the machine tool life, determining the number of parts that can be produced before the tool must be sharpened or replaced. The RP used in rust prevention determines the number of parts that rust and must be reworked or scrapped. This affects both the cost of the operation and the cost of the overall process.

Nonchemical inputs to any operation in the process can also carry chemical costs. For example, steam is used to heat water for the degreasing operation. Chemicals used to produce the steam not only contribute their TCO but also influence the efficiency of the steam production process.

Viewing the production process as a whole, it is easy to see how chemicals can have an enormous effect on product costs. The Chemical Beast tends to increase the life cycle costs of chemicals, increasing the cost not

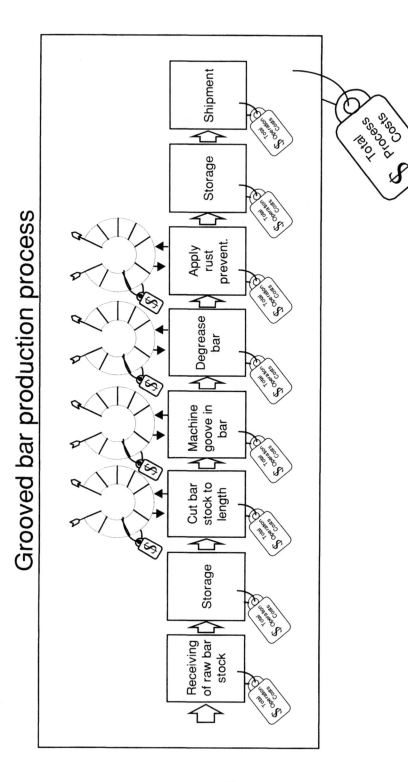

Figure 6-4. Chemical contributions to total process costs.

only of chemical inputs but also the cost of nonchemical inputs that are produced with chemicals. The Chemical Beast can inhibit the flow of ideas about how chemicals can improve operation efficiency, limiting a company's ability to remain cost competitive.

THE EFFECT OF CHEMICAL EXPERTISE

We have seen that chemicals can have a significant impact on product costs through the total cost of chemical ownership and through the effects of chemicals on operation efficiency. However, we have not yet examined the role of chemical expertise in the relationship between chemicals and product cost.

Fig. 6-3 demonstrated that chemicals affect operation costs in four ways. Chemical expertise plays a key role in each of these. Simply put, chemical expertise determines the time it takes to reduce chemical-driven operation costs. Even with minimal chemical expertise, most companies eventually make certain chemical improvements because the rest of their industry has done so. For some companies, this may be an acceptable way of doing business.

However, delaying available cost improvements represents lost profit opportunities and puts a company at a cost disadvantage compared to its competitors. In competitive markets, a cost disadvantage can have a significant effect on market share and may even determine the survival of the company.

The greater the chemical expertise available to the company, the more quickly it can make chemical improvements and reduce product costs, giving the company an important competitive advantage and increasing profits. As we demonstrate later in the book, companies who are able to tap their supplier's chemical expertise greatly increase their ability to reduce chemical-related costs.

IMPLICATIONS FOR THE MANAGER

From the perspective of production operations, chemical management can be viewed in a new light. Consider the following questions:

- How much does chemical TCO contribute the cost of each production operation?

- How much does chemical TCO contribute to the cost of nonchemical inputs such as steam and deionized water?
- How much do chemicals influence the efficiency of operations?
- What proportion of product cost is due to chemical TCO?

The impact of chemicals on product costs can be much greater than most managers perceive. The impact on overall business success can be even greater. It is to this subject that we turn in the next chapter.

7

Chemicals and Business Value

We wanted a supplier to look at the big picture, provide recommendations
to drive the total cost of the system down, and add greater value to the
business—not one who just tweaked the chemistry.

—A plant purchasing manager

The previous two chapters showed how chemicals can influence product
costs and how chemical improvements can provide a significant cost advantage over competitors. In this chapter, we broaden the perspective beyond
cost to include two other factors that are critical to business success: *quality*
and *capability*.

HOW BUSINESSES CREATE VALUE

Value is what customers pay for—and it's all they want to pay for.

The mission of business has always been to create value. The question is:
Value for whom? Prior to the quality revolution of the 1980s, the most common strategy in Western business was to create value for the shareholder.
A simple model of shareholder value is presented in Fig. 7-1. Shareholder
value can be thought of as the difference between total revenue and total
costs. The business generates revenues through the sales of its products,

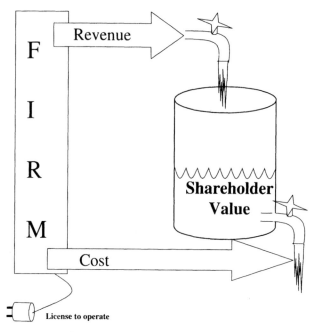

Figure 7-1. Traditional view of business value.

adding to the value tank of the shareholder. The business incurs costs through the development, production, and marketing of its products. These expenses drain value from the value tank. The level of value created for the shareholder is the difference between total revenue and total cost.

An additional factor that must be considered is the business's license to operate, denoted in Fig. 7-1 by the electrical plug. If a business complies with all applicable laws, it can plug in the business and operate. Failure to comply with the appropriate laws and regulations can result in pulling the plug on the business. The license to operate does not contribute to share- holder value, and the costs incurred to maintain the license drain value from the value tank. Nevertheless, a license to operate is necessary for operation.

In the 1980s and 1990s, managers came to recognize that focusing on creating *value for the customer* was a superior strategy for creating long-term shareholder value. In simple terms, customer value can be thought of as the difference between product quality and cost (sales price of the prod- uct). The business creates a product with specific qualities desired by the customer. This becomes an input to the customer's value tank. The price paid by the customer drains customer value but produces revenue for the business, adding value for the shareholder (Fig. 7-2). The business increases shareholder value by increasing total product quality and reducing total costs.

However, to sustain and improve customer value and, subsequently, business revenues, a business must continuously increase its capabilities. We define *capability* as the ability of the business to continuously increase value through higher product quality and lower operation cost. We address business capability in greater detail later in this chapter.

CHEMICALS AND BUSINESS VALUE

Improved chemical management practices can lead directly to increased business value, thus contributing to the competitive success of the company. Fig. 7-3 shows a simple business containing three production processes. Processes 1 and 2 produce components that are assembled in process 3. Operations in each process require chemicals, as indicated by the chemical life cycle above the operation. Improved chemical management increases

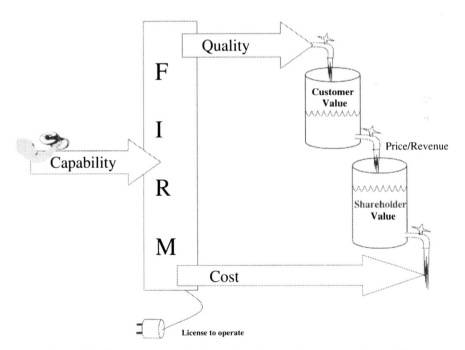

Figure 7-2. Business value is determined by quality, cost, and capability.

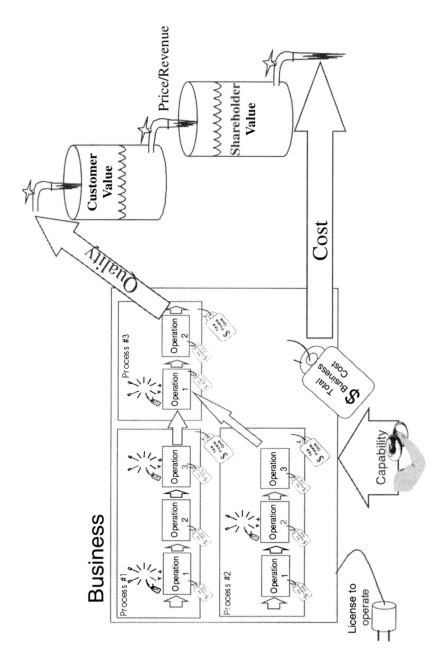

Figure 7-3. Chemicals and business value.

business value by reducing product costs, increasing product quality, and enhancing business capability.

Cost

The traditional practice of accounting ... end[s] up weakening American business.
—P.L. GRIECO AND M. PILACHOWSKI,
Activity-Based Costing: The Key to World-Class Performance

Controlling cost has always been important in business. Lower cost increases shareholder value by increasing the proportion of revenue that remains for shareholders (earnings). This allows a company to be more aggressive and competitive on product pricing, ultimately enhancing customer value.

It should not be surprising that cost control is a priority for companies when it comes to purchasing chemicals. However, as shown in previous chapters, efforts to reduce chemical costs by focusing solely on chemical purchase price are inadequate and even counter productive. For many companies, the chemical purchase price represents only a small portion of the total cost of ownership (TCO). The total cost of chemical ownership contributes to the cost of operations and, ultimately, the cost of products. In addition, many chemicals affect the efficiency of operations, further influencing product costs.

In chapter 6, we proposed that better management of chemicals could reduce costs in two basic ways. First, reducing the total cost of chemical ownership reduces the cost of chemical inputs to an operation. It also reduces the cost of nonchemical inputs that are produced or prepared with the use of chemicals. Second, chemicals can improve operation efficiency, reducing the amount of inputs needed to produce a unit of output. This is also true for ancillary operations used to prepare nonchemical inputs, reducing the overall cost of such inputs. Chemical management expertise plays a key role in identifying and implementing opportunities to reduce the total cost of chemical ownership and to improve the efficiency of operations.

Unfortunately, current cost accounting systems are rarely able to link product costs to the chemicals that produced those costs. Overhead allocation mechanisms often produce a distorted picture of the link between chemicals and product costs. To tame the Chemical Beast and use chemicals to increase business value, a manager must distinguish between the *real* chemical costs (Fig. 7-3) and the distorted picture of chemical costs provided by the company's cost accounting system.

Product Quality

Quality first, not profit first. . . . If you take care of the quality,
the profits will take care of themselves.
—MASAAKI IMAI, *Kaizen: The Key to Japan's Competitive Success*

Product quality is created in the production process and quality improvement can be achieved through process improvement. As a product moves through the production process, each step contributes to improved product quality and, ultimately, increased customer value. Chemicals contribute to this quality either directly, by becoming part of the product, or indirectly, by improving operation performance. As an example of direct contribution, switching from liquid to powder paint has increased finish durability for many products. However, chemicals can also improve product quality indirectly by improving the performance of operations. For example, a superior coolant can result in a higher-quality machined surface, and a superior cleaner can result in better surface bonding during electroplating.

Chemicals can also influence other, less tangible quality attributes. Though not functional attributes of the product, these quality attributes may be no less important to the customer and therefore contribute to the overall marketability of the product. The array of such quality attributes is limitless. We address two common quality attributes that can be strongly affected by chemicals and chemical management.

Uninterrupted product supply is a quality attribute valued by most customers, but it is especially important to business customers. Chemicals can disrupt production in many ways: a chemical spill, a dermatitis outbreak, a fire, equipment failure, boiler breakdowns, a malfunctioning wastewater treatment system, etc. Similarly, in serious cases, chemical-related regulatory violations can pull the plug on an entire process or plant. Many customers prefer to buy from suppliers who have a history of uninterrupted supply and a chemical management system designed to eliminate regulatory threats to production.

Reputation, both of product and manufacturer, is valued by many customers. Some business customers wish to avoid association with chemicals and suppliers who have a reputation for poor environmental or health and safety (EHS) performance. Many choose products and suppliers with a positive EHS image because it tends to enhance their own image (or is, at least, unlikely to detract from it). Improved chemical management can significantly improve EHS performance, enhancing a company's reputation with customers.

Capability

You need to . . . establish that elusive trait known as value.
Or, more properly, you need to establish the capabilities for
creating value more effectively than any other approach.
—JAMES F. MOORE, *The Death of Competition*

As important as quality and cost control are to creating value, enhancing a manufacturer's capability to create value is equally important. Managers who are constantly under pressure for short-term performance can easily overlook this point. Yet, to neglect the development of internal capability is to fall behind more proactive competitors.

Many factors contribute to capability. Among the most important are:

- *Adaptability*—the ability to adapt quickly to changing conditions. This requires flexibility not only in production but in all business processes. Factors that contribute to flexible processes include operational simplicity and efficiency. Thus, simplifying processes and reducing cycle times enhance adaptability. Chemical improvements can both simplify production operations and reduce operation cycle times.
- *Learning and Innovation*—the ability of an organization to learn or invent new products, processes, technologies, and strategies. Companies need to invest their learning and innovation resources to yield the greatest return—in areas of core competence. External sources of expertise in areas outside of one's core competence must be tapped. While managing chemicals better does not, in itself, lead to greater organizational learning or innovation, it is one of the greatest benefits from Shared Savings chemical management supply programs. As we will see, such supply programs provide impressive cost and quality benefits, but it is the added capability for learning and innovation that produces a quantum leap in competitive power.

- *Anticipation*—the ability to anticipate changes in customer needs and market conditions. This requires close connections with customers, suppliers, employees, and other sources of information about future trends. Again, while better management of chemicals does not, in itself, create better anticipation, Shared Savings chemical management programs do—and in an interesting way. As we will see, it allows a company's chemical suppliers to better anticipate problems or needs in the company's future production processes. This leads to lower future costs and higher future quality.

- *Intelligence*—useful knowledge about business operations. A thorough understanding of operations is essential to both control and improve production. In the following chapters, we demonstrate how Shared Savings chemical management produces better data systems and a freer flow of information than do traditional supply programs. The resulting intelligence about business operations gives the company greater capability to bring those operations under control and then improve them to reduce cost and increase quality.

IMPLICATIONS FOR THE MANAGER

How much value do chemicals, and your chemical suppliers, add to your business? Consider the following questions:

- How much shareholder value is drained by chemical-related costs?
- How many chemicals become part of your products, and how much do they influence product quality?
- How much do chemicals affect product quality by affecting the performance of production operations?
- How much do chemicals affect the reputation of your company and its products?
- How much do chemicals affect your company's ability to make rapid changes in products and processes?
- How much does your chemical supplier contribute to learning and innovation in your company?
- How well does your supplier anticipate your future chemical needs?
- How well does your company understand and optimize its chemical systems?

Some companies have recognized that chemicals, and their chemical suppliers, are far more important that previously believed. These companies found that their current chemical supply strategy was outdated and put them at a competitive disadvantage. In part 2 of this book, we explore alternative supply strategies that create greater value for the chemical user and help to tame the Chemical Beast.

PART 2

SHARED SAVINGS CHEMICAL MANAGEMENT

CHAPTER

8

Alternative Chemical Supply Strategies

When the goal is boosting profits by dramatically lowering costs,
a business should look first to what it buys.
—SHAWN TULLY

As we have seen, traditional chemical supply relationships contribute to the Chemical Beast. From misguided financial incentives to the division of chemical management responsibilities, traditional supply encourages chemical waste and inefficiency. Fortunately, many companies have recognized the limitations of the traditional chemical supply relationship and have developed superior alternatives that address them.

In this chapter, we construct a simple hierarchy of chemical supply alternatives. Movement to higher levels of the hierarchy can help contain the Chemical Beast and increase business value for the chemical user. To understand this hierarchy, it is necessary to understand how the alternative strategies differ.

Chemical supply relationships differ along a variety of dimensions. Five dimensions of greatest importance are:

1. *Fee structure*—what the supplier is paid for.
2. *Services*—the scope of services performed by the supplier.

3. *Problem solving*—the process used to solve problems and share information.
4. *Footprint*—the array of chemicals for which the supplier is responsible.
5. *Metrics*—how chemical and supplier performance is measured.

These dimensions are summarized in Fig. 8-1. The range of options for each dimension is presented, from a traditional supply relationship through the most advanced of today's Shared Savings chemical management relationships. Each dimension is discussed below.

FEE STRUCTURE

If you want suppliers to come in and do a good job in reducing chemical use when their traditional compensation was solely driven by margin and volume, you are going to have to change the way you compensate them. That message is clear.
—*A plant purchasing manager*

Fee structure is the method employed by the chemical user to compensate the chemical supplier. It defines a supplier's performance incentives. In the traditional supply relationship, the supplier is typically paid on a dollars-per-pound basis, creating an incentive that drives the volume conflict between buyer and supplier. Movement across the continuum of fee structures reduces this volume conflict and aligns the interests of chemical buyer and chemical supplier (see Fig. 8-2).

Dollars per Pound

The traditional chemical supply relationship is a dollars-per-pound relationship. Chemicals are sold on a dollar-per-pound (or gallon, or drum) basis. Chemical suppliers compete for business primarily on the basis of price alone and they profit through margins on the chemicals sold. The larger the sales volume, the greater the profit.

As we saw in chapter 4, this type of fee structure creates a volume conflict between the supplier and the user. If the user is paying the supplier for volume, it should not be surprising that that is exactly what the user gets.

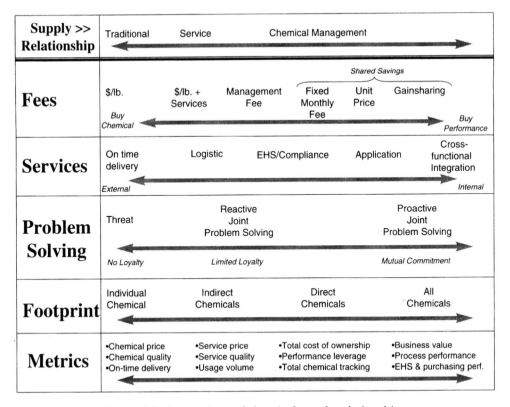

Figure 8-1. Dimensions of chemical supply relationships.

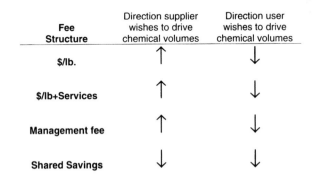

Fee Structure	Direction supplier wishes to drive chemical volumes	Direction user wishes to drive chemical volumes
$/lb.	↑	↓
$/lb+Services	↑	↓
Management fee	↑	↓
Shared Savings	↓	↓

Figure 8-2. Alternative supplier fee structures and volume conflict.

Dollars per Pound Plus Services

In a dollars-per-pound plus services fee structure, chemicals are still purchased on a dollar-per-pound basis, but related chemical services are a more prominent part of the package. The buyer pays for the supplier's services through higher per-pound prices. Chemical prices are competitive but generally not the lowest in the market. Suppliers profit from slightly higher margins but also hope to retain or expand customer accounts by providing better value-added services than their competitors.

Though this fee structure promotes a more cooperative relationship between chemical user and chemical supplier, the adversarial nature of the relationship is still intact. Supplier revenue is still linked directly to chemical volume, preserving the volume conflict between chemical user and chemical supplier.

Chemical Management Fee

In a *chemical management fee* structure, the supplier receives a fee to provide specified chemical management services. The cost of chemicals is passed through to the customer on a per-pound basis, but the supplier's management services are paid for as an itemized fee rather than higher chemical prices.

The most important difference between the dollars-per-pound plus services fee structure and the chemical management fee structure is that, in the latter, the supplier's management fee is independent of chemical volume. In some cases, the chemical user may establish mandatory performance specifications that require reductions in chemical use. Yet in this type of supply relationship suppliers receive no direct financial benefit from chemical reductions and still obtain most of their revenue from chemical sales. The volume conflict, though reduced, remains intact.

Nevertheless, this represents an important departure from traditional supply relationship fees. It implies that the supplier can provide value to the chemical user in ways that are independent of the chemicals supplied. In addition, it means the chemical user realizes that at least some aspects of chemical management are not part of the company's core business and is willing to pay the supplier for chemical knowledge and expertise. For the supplier, this is a major step toward becoming a chemical service company rather than just a chemical supply company.

Shared Savings

A Shared Savings fee program can be structured in several forms. The three most common forms—fixed fee, unit price, and gainsharing—are discussed in more detail in the next chapter. The underlying factor that is consistent across all Shared Savings fee structures is an incentive that aligns the interests of the chemical supplier with those of the chemical user. Just as the dollars-per-pound fee structure creates an inherently adversarial incentive, Shared Savings fee structures create an incentive that motivates both parties to continuously reduce chemical use and waste while continuously improving product and process quality. The Shared Savings fee is structured to allow supplier and the user to share the savings gained from these improvements.

In all Shared Savings fee structures, the chemical user no longer buys chemicals. Instead, the supplier is paid a fee to meet the chemical needs of the plant. Disconnecting supplier revenue from chemical volume has a dramatic effect. The supplier's revenues are fixed with respect to chemical volume, yet their costs vary directly with respect to the volume of chemical used. This creates an incentive for the supplier to reduce chemical use (and, thus, chemical costs) in order to increase profits (see Fig. 8-3). Chemical use reductions are achieved by improving the efficiency of chemical operations. This aligns the incentive of the chemical supplier with the interest of the chemical user—to drive chemical volumes down. This is the

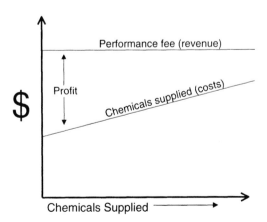

Figure 8-3. Shared Savings relationship—a supplier's incentive to decrease chemical volume.

opposite of the traditional chemical supply relationship, where the supplier's financial incentives reward increased chemical sales (see Fig. 4-2).

SERVICES

The chemical supplier's ability to deliver continuous improvement in value-added services for the bottom line—that's really what we are focused on right now.
—A plant purchasing manager

The array of potential supplier services, beginning with activities as basic as on-time delivery, is almost infinite. Most of these services can be grouped into three categories: logistic, environment, health, and safety (EHS)/compliance, and application.

Logistic Services

Logistic Services involve all functions related to the physical movement or containment of chemicals, including activities such as chemical procurement and payment, delivery, inventory management, internal distribution, and collection.

EHS/Compliance Services

EHS/compliance services involve all aspects of chemical-related regulatory compliance as well as activities intended to achieve the desired level to worker protection and environmental performance. Examples include chemical tracking for environmental reporting, container labeling, health and safety training, waste treatment, and proper disposal.

Application Services

Application services can include all aspects of maximizing the performance of a chemical, including testing and maintaining chemical quality, in-process chemical reclamation, and identifying and implementing new chemical application strategies. They may even involve the development of new chemical products. Many application services require significant technical expertise and close cooperation with the chemical user.

The Range of Services

Logistic, EHS/compliance, and application services can be offered by the supplier in any combination. Most commonly, chemical users begin with logistic services, followed by EHS/compliance and, finally, application services. Chemical users also tend, initially, to request services that are performed largely outside the plant, and later add services performed inside the plant. These trends are consistent with the need to establish trust in a supplier. Logistic services tend to carry less risk than EHS/compliance services and much less than application services. Similarly, services performed outside the plant tend to carry less risk than services inside the plant.

As illustrated in Fig. 8.1, the most advanced supplier service is what we call *cross-functional integration*. This involves provision of a full array of logistic, EHS/compliance, and application services and requires the supplier to help coordinate chemical management activities across the functional areas of the chemical user's plant.

PROBLEM SOLVING

> But it isn't enough to share an important objective.
> You also need a relationship for solving problems along the way.
> —JORDAN D. LEWIS, *Partnerships for Profit*

All production processes experience chemical problems on a periodic basis. All supply relationships have problems as well. How the chemical user chooses to manage the problem-solving process determines how much value is gained from the supply relationship.

In a traditional chemical supply relationship, problems associated with the chemical supplier are typically addressed through the use of threat. The buyer assumes that the threat of switching suppliers will be sufficient to correct any problems with the supplier or the supplier's chemicals. If the problem is not resolved, the chemical user brings in another chemical supplier.

Some chemical users place greater value on the expertise of their supplier. They may seek the supplier's input regularly to resolve chemical-related problems in the plant. This creates a limited amount of loyalty on the part of the chemical user. The supplier is usually confronted with a price, quality, or performance problem and given an opportunity to correct it. Only if the problem is not corrected, or if it recurs frequently, will the chemical user resort to threat.

At the extreme, chemical users may seek suppliers primarily for their proactive problem-solving expertise and ability. These chemical users recognize that joint problem solving is essential for long-term cost reductions and performance improvements. Supplier personnel work closely, usually daily, with chemical user personnel. In many cases, the supplier discovers chemical problems first and brings them to the attention of the chemical user's staff. As trust is built, so is a spirit of mutual commitment to making the relationship work and to work together to resolve even the most difficult problems.

FOOTPRINT

The chemical footprint of a supplier is defined by the number of chemicals and type of chemicals for which the supplier is responsible in a given plant or process. In a traditional supply relationship, a supplier may be responsible for only one chemical. In fact, in some companies, the same chemical may be supplied by a number of suppliers.

Many companies have begun to consolidate chemical suppliers. This is perceived to be riskier than using multiple suppliers because a supply failure has a greater impact on production. However, it also opens a multitude of opportunities for the chemical user to obtain cost and performance improvements. Having fewer suppliers allows the chemical user to develop closer working relationships with those that are retained. For their part, suppliers have a better opportunity to understand the plant's product, process, and problems. In some cases, a company may designate a Tier 1 supplier with management responsibilities over Tier 2 suppliers.

METRICS

In short, you become what you measure.
—D.A. RIGGS AND S.L. ROBBINS,
The Executive's Guide to Supply Management Strategies

Appropriate metrics are essential for a company to monitor and improve the success of chemical management activities. The scope of metrics in a traditional supply relationship is limited. Price may be the primary measure of success—the lower the price, the greater the success. Timely delivery and chemical quality may also be monitored to document performance and identify potential problems.

More advanced metrics, including service quality and service price, are used, particularly in supply programs that cover a broad range of supplier services. Whether a buyer is receiving services through higher product prices or paying for them directly, price and quality are important measures of the chemical management program's success. Because stabilizing or reducing chemical volume is often a priority for the buyer, chemicals may be monitored as well. This monitoring offers additional benefits for regulatory reporting, as much of the data collected is required to complete mandatory annual reports.

Measurement of chemical quality and delivery remains important but is no longer useful in evaluating the chemical management program because the supplier is expected to meet or exceed the user's expectation on a continual basis. *Any* problem with quality or delivery is considered serious and expected to be corrected immediately. In some cases, our research found that the supplier was required to reimburse the chemical user for costs or losses related to quality or delivery problems.

As noted in chapter 2, the total cost of ownership (TCO) can be significantly larger than the chemical purchase price. Total cost of chemical ownership is the actual cost that companies need to minimize. Some companies are attempting to measure the total cost of chemical ownership as a better measure of chemical management performance than mere price.

SUPPLY RELATIONSHIP HIERARCHY

We suggest to the buyer that a new paradigm would allow us to bring savings to the arrangement that would greatly exceed typical pricing concessions.
—C.C. POIRIER AND W.F. HOUSER,
Business Partnering for Continuous Improvement

Fig. 8-1 suggests that the combination of options for fees, services, problem solving, footprint, and metrics can produce an almost infinite array of supplier relationships. We find it useful to group these possible relationships into four hierarchical categories (see Fig. 8-4):

1. Transactional (traditional) relationships
2. Service relationships
3. Limited chemical management relationships
4. Shared Savings chemical management relationships

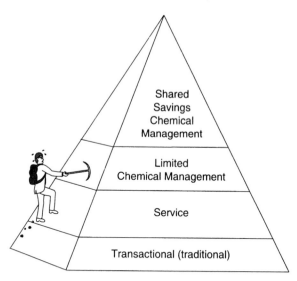

Figure 8-4. The hierarchy of supply relationships.

Transactional Relationships

Traditional chemical supply relationships are characterized by an almost total focus on the transaction process—exchanging chemicals for money. The chemical user's primary concern is finding the supplier with the lowest price and using any type of competitive leverage to keep prices low. Supplier profits are driven by volume, producing an inherent conflict of interest between the chemical user and chemical supplier.

Other than low price, a specified level of chemical quality, and on-time delivery, little is expected of the suppliers. Performance metrics are often limited to these three areas. Because supplier success is defined through short-term measures (price, quality, delivery), the threat of switching suppliers is often used as leverage to correct performance problems. This short-term focus limits the supplier's interest in solving problems that require long-term investment.

Transactional supply relationships can apply to both large and small chemical footprints, though they are more commonly used with small footprints. The supplier typically provides only one or just a few chemicals. Some chemicals may even have multiple suppliers. However, many companies consolidate suppliers without changing the traditional supply relationship. These suppliers may provide an array of chemicals, yet price, quality, and delivery remain the primary focus of the relationship.

Service Relationships

Service relationships offer increased value for the chemical user over a transactional strategy. Chemicals are still purchased on a volume basis, but related chemical management services are a more significant component of the supply relationship (a dollars-per-pound plus services fee structure).

In fact, the scope of services offered by a supplier can be a determining factor in supplier selection. Services can cover a range of options but generally emphasize logistic services, particularly outside of the plant—such as improved chemical packaging, just-in-time (JIT) deliveries, electronic data interface (EDI) ordering, etc. Occasionally, EHS/compliance and even application services may be included to supplement logistical services. For example, some suppliers offer labeling and signage compliance both inside and outside of the plant. Many suppliers include in-plant testing of their chemicals and even advice on usage.

Limited Chemical Management

The boundary between service and chemical management relationships is not clearly defined. The most important difference is problem solving. In chemical management programs, the suppliers are valued largely for their chemical and management expertise. Thus, a supplier's participation in joint problem solving is a central component of the relationship. The supplier participates in the development of solutions through the provision of chemical expertise as well as management expertise and experience. This contrasts with service and transactional relationships, in which the supplier responds largely to the demands of the buyer.

In limited chemical management programs, fee structures include dollars-per-pound plus services and management fees. Supplier services tend to be heavily weighted toward logistic (particularly inventory management) and EHS/compliance activities, though in some cases the full range of services may be provided. Services are often provided both inside and outside the buyer's plant. In many cases, the supplier assigns one or more of full-time personnel to the plant. Occasionally, these limited relationships can actually be quite extensive. We are aware of one program based on a dollars-per-pound plus services fee structure that uses almost the full range of supplier services and is characterized by a long-standing history of mutual commitment between buyer and supplier.

Shared Savings Chemical Supply Relationships

A chemical management relationship becomes a Shared Savings chemical management relationship when the supplier fee is structured to align the interests of chemical user and chemical supplier. Current Shared Savings programs use one or more of these fee structures: fixed fees, unit prices, and gainsharing. This produces profound changes in the relationship. Aligning the financial interests of buyer and supplier to reduce chemical volume and increase value for the ultimate consumer dramatically improves cooperation, continuous improvement, and innovation.

Shared Savings programs create risks for both parties, but they also generate far greater opportunities than do traditional relationships for enhancing financial returns. Programs typically include comprehensive supplier services, proactive problem solving, a large chemical footprint, and more extensive metrics for monitoring TCO, performance leverage, and chemical tracking.

In the next chapter we outline the elements of Shared Savings chemical management. Then, in chapter 10 we contrast Shared Savings with traditional chemical supply relationships and explain how Shared Savings produces dramatically greater business value.

IMPLICATIONS FOR THE MANAGER

How would you characterize the current relationship with your chemical suppliers? Using Fig. 8-1, describe your current supply program in terms of:

- Fee structure
- Services
- Problem solving
- Footprint
- Metrics

Would you characterize current chemical supply as traditional, service, limited chemical management, or Shared Savings chemical management? What problems are caused by the supply relationship as it is currently structured? What benefits might result from moving to a higher level—toward Shared Savings?

In the next chapter, we take a closer look at the essential elements of a Shared Savings chemical management relationship.

CHAPTER

9

Elements of Shared Savings Chemical Management

Now, all our arrows are heading in the same direction.
—ANDERS KAUSERUD, *on-site manager,*
Castrol Industrial North America

In this chapter we summarize the essential elements of a Shared Savings chemical management program. In later chapters we provide more detail on aspects of Shared Savings such as the contract, pricing, appropriate metrics, and implementation.

FEE STRUCTURE

In all Shared Savings fee structures, the chemical user no longer buys chemicals. Though the chemicals typically become the property of the user at the point that the chemical is used in the process, payment to the supplier is not tied to chemical volume. Instead, the supplier receives a fee in exchange for meeting certain performance expectations (explained below). Shared Savings fee structure is discussed in more detail in chapter 24.

The three most common Shared Savings fee structures are explained below.

Fixed Fees

Under a *fixed fee* structure, suppliers are typically paid a fixed monthly fee. In return, the supplier agrees to meet the performance expectations negotiated with the plant. Historical chemical usage and costs usually determine the monthly fee, which may be higher than historical costs to compensate for additional services, or lower, if the supplier is able to offer price concessions.

As discussed in the previous chapter, the fixed fee creates a strong incentive for the supplier to reduce chemical usage and waste. In other words, the supplier increases profits by decreasing chemical volumes, as opposed to traditional supply relationships in which supplier profits are increased by selling more chemicals. Ultimately, some of these savings must be shared with the chemical user so both parties have an incentive to mutually pursue further cost reductions. One strategy is to rebate some of the savings to the chemical user. Another strategy is for the chemical user to reduce the fixed fee to reflect the supplier's lower chemical costs.

In the case histories provided in the next section of the book, chemical management programs at both Navistar and the General Motors (GM) Electro-Motive Division use a fixed-fee pricing structure. Both include a mechanism for rebating a portion of large supplier cost savings back to the chemical user. They also include provisions for periodically adjusting the fixed fee as chemical usage declines.

Unit Prices

A *unit price* is a fee paid to the chemical supplier for every unit of product produced by the chemical user. For example, the supplier might be paid $5 for each automobile or washing machine produced by the plant. Variations of the unit price may link supplier fees to an intermediate step in the production process. For example, a supplier of paint detackification chemicals might be paid a fee per gallon of paint sprayed in the paint shop. A supplier of boiler water treatment chemicals might be paid per million pounds of steam produced.

Unit prices are similar to fixed fees in that they are independent of the volume of chemicals supplied. This eliminates the volume conflict and aligns the incentives of chemical user and chemical supplier to reduce chemical volume. Unit prices can be coupled with rebates to share large cost savings or reduced to reflect lower supplier costs.

In the case histories, Ford, Chrysler, and the GM Truck and Bus plant all use unit prices. Prices were derived from historical chemical costs and production volumes, though some modification was made to reflect additional supplier services as well as supplier discounts.

Gainsharing

Gainsharing is a mechanism by which a buyer directly shares cost savings with the supplier. If a supplier's idea or innovation generates savings for the buyer, a gainsharing agreement distributes those savings between both parties.

Gainsharing agreements are typically used in combination with unit prices or fixed fees, but they may also be used in conjunction with management fees or even a dollars-per-pound plus services fee structure (see chapter 8). The value of gainsharing is that it strengthens the alignment of the buyer's and supplier's financial interests. Because gainsharing can be applied to any savings, even those unrelated to chemicals, it expands the potential benefits of the supply relationship.

To illustrate the value of gainsharing, consider two potential innovations. The first innovation is a new coating that allows paint shop walls to be cleaned with soap and water instead of solvent, reducing solvent use by 100,000 gallons per year. The second innovation is a new corrugated cardboard compactor that saves the chemical user $30,000 per year in cardboard storage and handling and generates recycling revenue.

Under a unit-price or fixed-fee structure, the supplier has a strong incentive to implement the first innovation, as it would dramatically lower the supplier's costs. However, the supplier has little incentive to implement the second innovation, as it would not benefit from the savings. With gainsharing, however, a portion of the $30,000 annual savings would be shared with the supplier, creating an incentive to bring this and future innovations to the attention of the chemical user.

While some suppliers argue that they would bring such innovations to the attention of the buyer simply to promote customer loyalty, our research suggests that goodwill alone does not provide the level of incentive possible under a gainsharing program. The level of innovative effort is higher when the supplier's personnel know that they can directly improve their financial performance measures every time the buyer implements one of their cost savings innovations.

Though gainsharing has been used at a number of plants with Shared

Savings chemical management programs, the five plants that we profile in this book did not have gainsharing experience. It had just been implemented at the Navistar plant at the time of our study. The GM Truck and Bus plant and the Ford Chicago assembly plant were exploring how to build gainsharing into their programs.

SERVICES

Shared Savings chemical management programs typically include the full range of logistic, environment, health, and safety (EHS)/compliance, and application services. Many also have achieved a level of cross-functional integration. In the plants that use chemical suppliers to facilitate cross-functional integration, we found far greater control of the Chemical Beast. In those plants that utilize suppliers as cross-functional coordinators, supplier representatives in the company become more fully integrated into the workforce and achieve a higher level of success.

Supplier services are typically specified in a set of *performance expectations* that clearly define what the chemicals and chemical supplier are expected to do. These expectations go well beyond the type and volume of chemicals required. They can even go beyond the traditional definition of services to include product quality expectations, equipment operating characteristics, tool life expectations, corrosion requirements, etc.

Supplier fees are often explicitly linked to performance expectations. In some cases, fees are not paid if performance expectations are not met. For example, the Chrysler Belvidere assembly plant utilizes a *unit-price* payment strategy called the *Pay-as-Painted program*. Chrysler's paint supplier, PPG, is paid a predetermined fee for each vehicle that leaves the paint shop with a finish meeting Chrysler's performance expectation for finish quality. If the vehicle does not meet Chrysler's specifications, PPG does not get paid. This connection between supplier revenue and chemical performance focuses the supplier on assuring the performance of their chemicals rather than simply supplying them.

One or more supplier personnel typically work full time at the chemical user's facility in order to meet performance expectations. These people work closely with plant personnel, from production workers to the plant manager. Supplier personnel typically hold degrees in chemical engineering or another technical field, have a number of years of industrial experience, and are *not* part of the supplier's traditional sales force.

PROBLEM SOLVING

One of the most important elements of a Shared Savings chemical management program is its emphasis on proactive, joint problem solving. Suppliers are expected to seek out chemical-related problems and to work jointly with plant personnel to solve them. All of the plants we studied had formed standing chemical management teams to anticipate and solve problems. These teams were composed of staff from all areas of the plant affected by chemicals, including Purchasing, EHS, Engineering, Maintenance, Production, and unions. Chemical supplier personnel were vital to these teams, often taking a leadership role.

Personnel we interviewed at these plants continuously referred to the chemical management teams as the means by which improvements were made. The GM Electro-Motive Division facility provides a particularly clear example. The plant's chemical management team worked out a ten-step process by which all proposed changes were reviewed by personnel likely to be affected by the changes. This included all the regular members of the chemical management team and personnel with an interest in the specific change being considered, such as machine operators or maintenance staff. The purpose of this approach was not only to consider all of the possible implications of the change but to be sure that no one felt excluded from the process. This assured smooth implementation of the changes. While the ten-step process required time to reach consensus, it dramatically accelerated the rate at which chemical improvements were instituted at the plant.

FOOTPRINT

Shared Savings chemical management requires dramatic consolidation of chemical suppliers. In most cases, the plant has a single Tier 1 supplier for almost all of the plant's chemicals or Tier 1 suppliers for at least its major chemical groups—coolants, cleaners, and oils, paints, solvents, water treatment chemicals, etc.

The need to consolidate is driven by two primary factors. First, placing chemical performance responsibilities in the hands of a single supplier eliminates the problem of one supplier blaming another for chemical performance problems. It also simplifies problem solving by giving plant personnel a single point of contact for all chemical-related problems. Second, consolidation is necessary to produce a contract of sufficient magnitude to support the on-site expertise provided by the supplier.

In some programs, most of the chemicals in the contract footprint are produced or distributed by the supplier. However, this is not always the case. The Tier 1 supplier may be responsible for an array of chemicals outside of its chemical specialty. For example, a coolant supplier may hold the Tier 1 position in a plant, but the contract may cover solvents and water treatment chemicals as well. In this situation, Tier 2 suppliers are contracted for these areas. The Tier 2 suppliers may provide part-time or full-time on-site personnel under the direction of the Tier 1 supplier.

In some cases, the Tier 1 supplier may even contract with competitors for Tier 2 chemical supply. For example, the coolant supplier holding the Tier 1 position may buy many of the plant's coolants from other coolant suppliers. This is common when the plant is pleased with its current chemicals but desires to consolidate chemical services under one supplier. The Tier 1 supplier can substitute its own chemicals only as it demonstrates that they are equal or superior to existing chemicals.

METRICS

Metrics for Shared Savings chemical management programs go well beyond price, chemical quality, and on-time delivery. Because the purposes of Shared Savings programs are to reduce the total cost of chemical ownership, improve product quality, and enhance the plant's capabilities, metrics should track these objectives. Unfortunately, actual metrics used in Shared Savings programs fall far short of providing adequate information in these areas. Current accounting systems, even activity-based costing systems such as those in use at Chrysler, are inadequate to fully measure the success of Shared Savings programs.

However, a number of valuable metrics are in use at many plants with Shared Savings programs. Most important is *chemical tracking*. Prior to implementing a chemical management program, most plants had relatively poor chemical tracking systems. They could provide rough estimates of overall annual chemical use, but the accuracy of these estimates were questionable and little information was available on usage by department, much less by operation or piece of equipment. This changed rapidly with the implementation of Shared Savings. Because supplier profit is dependent on careful control of chemical use, suppliers helped implement extensive chemical tracking systems. In some cases, chemical consumption is monitored down to the individual piece of equipment.

Improved chemical tracking allows companies to use valuable chemical efficiency metrics. Simply dividing chemical usage by product output provides a powerful means of tracking and encouraging chemical efficiency. Such metrics can be used at the plant, departmental, process, or equipment level.

IMPLICATIONS FOR THE MANAGER

Could a Shared Savings chemical management program work in your plant?

What would change at your plant if you had a single chemical supplier to work with, and that supplier made *more profit* when your plant used *less chemical*? What would change if your chemical supplier was bringing you more efficiency improvement ideas than your staff had time to evaluate? What would change if the supplier was performing most of the chemical management activities, from reordering to inventory to disposal?

Many plants have found that these changes produce dramatic reductions in costs as well as improvements in quality and plant capability. In short, they are taming the Chemical Beast. How does Shared Savings chemical management work? This is the question we address in the next chapter.

CHAPTER

10

Why Shared Savings Works

Shared Savings works because it creates greater
business value than other supply strategies.

Traditional chemical supply creates a cycle of ever-growing chemical problems. It begins with the adversarial interests, which lead to a focus on chemical transaction, which leads both parties to protect their own interests—
reinforcing their adversarial interests. As the cycle continues, the Chemical
Beast grows.

Companies that have implemented Shared Savings chemical management strategies, however, operate under a different model (Fig. 10-1). In
this chapter, we explain how the steps in the Shared Savings model create
a self-perpetuating cycle that contributes to improved chemical control,
lower chemical costs, and greater business value. Finally, we explore how
Shared Savings has been used to overcome the other key factors that drive
the Chemical Beast.

THE PRINCIPLE: ALIGNED INTERESTS

Instead of a vendor wanting to sell more product to a customer, now the vendor
is working with the customer to optimize and reduce the excess in the system.
—LARRY PETTY, *PPG On-Site Manager, Chrysler Belvidere Assembly Plant*

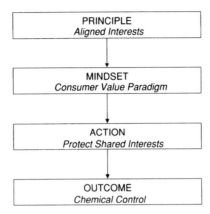

Figure 10-1. Shared Savings supply relationships.

The traditional chemical supply relationship is based on the principle that the chemical supplier and chemical user have inherently adversarial interests. Suppliers attempt to sell more chemicals or increase prices while users attempt to reduce chemical use and lower prices. In contrast, the Shared Savings supply relationship is based on the principle that it is possible—and *necessary*—to align the financial interests of chemical supplier and chemical user. One way this is done is to link supplier revenue to *value, not volume.* With a fixed fee or unit price structure, profitability is tied to performance rather than the volume of chemical supplied to the plant. In addition, the supplier is directly rewarded for continuous improvement by sharing in the resultant savings to the chemical user. The supplier makes more money only by creating greater value for the chemical user.

THE MINDSET: CONSUMER VALUE

For those who still prefer the traditional purchasing concept—forcing as much of the cost of doing business on suppliers as possible—we point out that all costs are eventually borne by the customer.
—C.C. POIRIER AND W.F. HOUSER,
Business Partnering for Continuous Improvement

All companies must create value for the ultimate consumer—who purchases the final product—if they are to survive in a competitive market. The greater the value of the final product, the greater a company's sales and market share. Consumer value in a product, as discussed in chapter 7, is defined as having features that produce the greatest consumer satisfaction at the lowest price. Selecting from a variety of competing products, a consumer purchases the product with the highest value. Thus, market share is won by the product that offers the greatest value to the ultimate consumer. Companies must continuously strive to increase consumer value if they intend to maintain or increase their market share.

Managers who recognize that consumer value is their real bottom line are following a *consumer value paradigm.* This is the antithesis of the *transaction paradigm,* where value is believed to come from the transaction process of exchanging chemicals for money (see chapter 4). From a Shared Savings perspective, however, value is created only when the chemicals meet or exceed the performance needs of the user.

Moreover, the product manufacturer is not the only one who determines ultimate consumer value. All of the upstream suppliers and all of the downstream companies that process, transport, or market the product contribute to ultimate consumer value as well. In fact, the entire process of creating value for the ultimate consumer is best viewed as a chain, often called a value chain. A simple value chain is illustrated in Fig. 10-2. (For purposes of illustrating the importance of the supply relationship, this chain contains only one supplier, one manufacturer, and one distributor. Most value chains are much longer and far more complex.)

Value is added to the product at each step in the chain. If supplier and manufacturer are able to improve processes or reduce costs anywhere in the value chain, they increase product value and profit through increased consumer sales. On the other hand, if they are unable to increase value as quickly as competing value chains, both ultimately lose profit though lost sales.

To succeed, the supplier and manufacturer must *eliminate* costs, not simply shift them back and forth along the same value chain. The fortunes

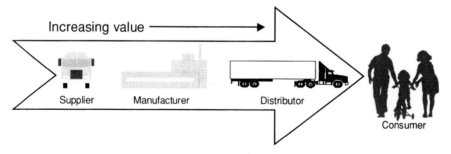

Figure 10-2. The value chain.

of all companies in a value chain rise and fall together with consumer value. Unfortunately, a transactional supply relationship pits the supplier against the manufacturer, placing them in competition. It creates the illusion that simply shifting costs to others in the value chain can increase long-term profits. However, cost shifting creates no additional value for the consumer.

> Costs are moved from the buyer's profit and loss statement to the supplier's. . . . The margins on the business will be shaved, leaving the incumbent supplier to find other hidden ways to offset the absorbed cost. . . . Absorbing cost within a buyer-seller network instead of working together to eliminate those costs is not a true advantage to the buyer. It puts the network at a competitive disadvantage relative to a more resourceful arrangement. (Poirier and Houser 1993)

Your true competitors are not other companies in your value chain but companies in value chains that compete for your market.

Under the consumer value paradigm, focusing on the chemical purchase price no longer makes sense. While chemical cost remains an important part of value, it is *total cost* (including hidden costs), not purchase price, that matters. Similarly, the loading dock no longer is an appropriate interface for transferring chemical responsibilities from supplier and user. Instead, each party assumes the chemical management tasks and responsibilities consistent with their respective core competence or expertise regardless of where in the chemical supply process these skills are required. In addition, because long-term customer value requires long-term investment, trust must be developed between supplier and the manufacturer. Each party must be willing to invest in long term, value-added projects.

ACTION: PROTECT YOUR SHARED INTERESTS

> We are seeing a major effort on [our supplier's] part to
> think of ways to lower their cost and save us money.
> —DAN UHLE, *Environmental Engineer, Ford Chicago Assembly Plant*

The consumer value paradigm, driven by the financial incentives of the Shared Savings relationship, leads to joint action by the chemical user and chemical supplier to increase ultimate consumer value. In particular, it leads to greater sharing of information and a willingness to invest in the future. The byproduct of this joint effort is increased control of the Chemical Beast.

The case histories presented in chapters 15–19 include many illustrations of this cooperative behavior reducing the impact of the Chemical Beast. Below, we highlight a few examples from the case histories and from other plants we have studied. Contrast them with the examples of conflicting interest in chapter 4.

- At a large metalworking facility, the supplier bought and installed its own coolant reclamation equipment. Field tests convinced the manufacturer that the reclaimed fluid was equal in quality to virgin coolant, while the reduction in waste disposal costs produced significant saving. Because the supplier paid for coolant supplies, the cost of the reclamation equipment was paid back by the savings realized from reduced coolant volumes.
- The chemical supplier to a large plant invested over $50,000 in consulting work to improve the treatment of paint booth wastewater. The investment reduced chemical use (saving the supplier money) and provided benefits to the chemical user in terms of reduced process downtime, reduced personnel overtime, and reduced waste disposal costs.
- The paint supplier for a large auto assembly plant invested research and development resources in creating a new process for repairing vehicles that experienced paint damage during assembly. The new process reduced volatile organic compound (VOC) emissions and costs while improving paint quality.
- Several suppliers developed advanced chemical tracking systems that result in improved chemical use efficiency and reduced material and labor costs. Implementing internal chemical tracking and management systems have given these suppliers significant market advantage over their chemical competitors.

THE SHARED SAVINGS CYCLE

This program provides much more help in maintaining control than I've ever had.
—DAN UHLE, *Environmental Engineer, Ford Chicago Assembly Plant*

The traditional chemical supply relationship produces a cycle of distrust leading to growth of the Chemical Beast. It is important to note that the distrust is a byproduct of the adversarial interests created through the

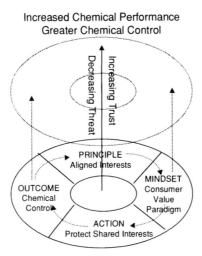

Figure 10-3. Shared Savings cycle.

transactional relationship, not a natural characteristic of either the suppliers or the users. In our research, we found that the Shared Savings chemical supply relationship creates a cycle of increasing trust between the supplier and the user, leading to greater chemical control and lower overall chemical costs (see Fig. 10-3).

In all plants we studied, including the five featured in this book, comprehensive Shared Savings programs grew from small pilot projects. As these limited programs succeeded, trust increased and the programs expanded. Increased trust between the supplier and the user does not just happen—it is earned by both parties a little bit at a time.

However, increased trust does not make the threat of competition irrelevant. For the supplier, the threat of being replaced by a competing supplier no longer is the primary motivation for performance. It moves into the background but does not completely disappear. The primary driver of the relationship for both parties becomes the performance of the relationship, which includes increasing chemical control, decreasing chemical costs, and increasing ultimate consumer value.

SHARED SAVINGS REDUCES THE OTHER CAUSES OF THE CHEMICAL BEAST

While the chemical supply relationship is critical to the size of the Chemical Beast, a number of other factors are also important, as discussed in chapter 3. Shared Savings programs help to overcome these factors as well.

Focus on Core Business

When people focus on core business, an interesting thing happens. We call it *entrepreneurial energy*. This is a change in the behavior of both supplier and user personnel as their activities move closer to the core business of their respective companies. The results are a greater level of energy, more focus, and higher creativity. Job satisfaction and success increase.

We observed this phenomenon in the chemical supplier's field staff. The performance success for these individuals was directly related to their employer's (the chemical supplier's) core business and contributed directly to professional recognition and advancement within their company. Supplier personnel could see the relationship between excellence in chemical management and a rewarding career. Furthermore, because these individuals worked for the supplier, they were not diverted to other activities in the plant. This was a frequent problem in the plants before chemical management was implemented, as chemical staff were sent to other areas of the plant, or even to other plants, whenever there were personnel shortages. Under Shared Savings, the supplier's chemical management personnel remained focused on chemical management responsibilities. As a result, chemical management activities were far more organized and controlled than they had been in the past. These factors contributed to the intense effort and high level of success of supplier personnel evident in the five case histories we present.

Similarly, manufacturers' personnel develop greater entrepreneurial energy as their responsibilities were directed closer to the core business of their employer. Companies using Shared Savings do not ignore the chemical management expertise in their own personnel. On the contrary, we found that chemical management personnel in most companies believed they become more knowledgeable about the chemicals used in their production processes. The most successful companies tapped the expertise of their suppliers to create and maintain expertise in their own production personnel. These company experts could then devote more time to identifying and resolving chemical application problems on a proactive basis instead of fighting fires and struggling with day-to-day chemical related tasks.

For example, environment, health, and safety (EHS) managers often took on a greater role in product design and consulting on the impact of chemicals in new production processes.

Untangle the Chemical Web

Under traditional chemical supply programs, the supplier sometimes benefits from complicated and confusing chemical systems. The less the chemical user understands, the less able they are to improve chemical use efficiency. Under Shared Savings chemical management programs, however, both chemical supplier and chemical user benefit from simple and understandable chemical systems.

This is why two of the most important goals in Shared Savings programs is to better understand the plant's chemical processes and then to consolidate chemicals. Chemical consolidation not only simplifies purchasing and inventory management, it reduces the complications in chemical applications, simplifying the production process. Under Shared Savings, it is not surprising to find companies making chemical consolidations of as much as 50–90%. For example, some plants that used twenty to thirty different coolants are now able to perform better using just two or three carefully chosen coolants.

Stay Ahead of Chemical Technology and Regulation

When chemical management expertise can be obtained from a chemical supplier (whose core business is chemicals), a company can redirect more of its resources to develop its own core business. Many chemical supply companies are expert in chemical procurement, delivery, inventory management, production applications, waste minimization, reclamation and recycling, and regulatory compliance. They also possess resources for chemical research and development that far exceed anything available within most chemical-using companies. For chemical suppliers, this is their core business, so they are able allocate the costs of developing chemical expertise and capability over a large volume of chemicals and customers, thus reducing the impact of their development expenses.

In the previous chapter, we mentioned two of the many examples of this we observed in our research. In one case, the supplier had developed a high level of EHS expertise on its chemicals from its internal EHS program. In another, the supplier had extensive research and development (R&D) capabilities, which were used for their own engineering and market development needs. In both cases, the marginal cost of extending this expertise to customers was small compared to the cost of customers developing the expertise on their own. Once the chemical supply relationships were structured to promote the sharing of this expertise, the chemical users experienced significant cost savings and improvements in chemical performance—in short, a tamer Chemical Beast.

The Right Information

Shared Savings programs thrive on high-quality information. The better the information, the greater the profits for both chemical supplier and chemical user. As a result, getting the right information is a high priority in Shared Savings programs. Information systems tend to focus on understanding the chemical needs of production processes, controlling those processes, managing chemical logistics, monitoring costs, and tracking improvements.

Efforts to understand the chemical needs of production processes begin even before a Shared Savings contract is signed, as proper contract terms require at least a basic understanding of existing chemical applications and costs. However, the most significant chemical related improvements to the information systems often occur once a Shared Savings program is implemented. Accurate chemical volume and cost data are essential to ensure the supplier's ability to assess potential profitability because the supplier's ability to improve efficiency and reduce costs can make a significant difference between profit and loss on the account. Development and implementation of an efficient and effective chemical tracking system is typically a high priority for both chemical user and chemical supplier.

Accurate cost accounting systems are often developed more slowly, in large part because the overall plant cost accounting system is controlled by the chemical user and the supplier has limited input. In response, suppliers may develop ad hoc cost accounting practices and systems, maintaining chemical cost data in addition to the data maintained for the plant's cost accounting system.

Coordinating Chemicals Across Departments

Overcoming the organizational barriers to cross-functional coordination of chemical management is an important aspect of a successful Shared Savings program. In all of the plants we studied, Shared Savings programs were coordinated through one or more cross-functional teams; typically, membership included supplier personnel as well as plant personnel from departments such as EHS, manufacturing, production control, maintenance, finance, purchasing, and unions.

Managers at the GM Truck and Bus facility believed that cross-functional coordination was so important to a successful Shared Savings program that it was included as an explicit part of the contractual agreement with the chemical supplier, requiring the supplier to "coordinate chemical usage with all affected plant departments." At Chrysler, the exceptional performance of the cross-functional chemical management program team is considered one of the company's greatest competitive advantages in the application of chemical management. At GM's Electro-Motive Division, the chemical management committee is the heart of the chemical management program. At Navistar, while cross-functional teams existed before Shared Savings, the inclusion of the on-site supplier representative on the team improved team effectiveness dramatically. To quote the Navistar environmental manager for the plant, "A decade ago, the coolant committee just didn't function that well. But under this new program, it's tied us all together—Production, Purchasing, Maintenance, and others. It's been a very good change."

IMPLICATIONS FOR THE MANAGER

One of the most informative quotations from our interviews came from Bob Conrad of Chrysler Corporation, who reflected on his experience with Shared Savings, "It requires a whole different kind of thinking."

Perhaps no other quotation better summarizes the barrier that companies face when considering Shared Savings. When viewed from the perspective of the traditional chemical supply program, Shared Savings just doesn't make sense. It seems to fly in the face of the principles that have guided supply programs from the start of the industrial revolution.

We were skeptical as well, until we saw Shared Savings at work in plant after plant. In these programs, both chemical user and chemical supplier seemed to be operating under a different set of rules. The never-ending cycle of chemical problems that plagued other plants was absent at these. Chemical systems were getting simpler, not more complicated; chemical costs were going down, not up; and chemical waste—with its associated compliance headaches—was declining rapidly.

It required "a whole different kind of thinking," but once the change was made, these plants would never go back.

CHAPTER

Purchasing and
Chemical Management

We keep missing the point! This isn't about buying "stuff" or checking to make sure we paid the right amount. This is about how suppliers can get us to the market faster and how we can be better when we get there.

—D.A. RIGGS AND S.L. ROBBINS,
The Executive's Guide to Supply Management Strategies

For the chemical user, chemical management has two sides: supply management and technical management. Improving control over the purchasing of chemicals can produce enormous benefits. Similarly, improving the application and performance of chemicals in the production process can have great value. This is why purchasing and environment, health, and safety (EHS) are often the two departments most actively involved in chemical management programs. Both departments work closely with production, engineering, maintenance, and other affected departments to facilitate the chemical management process. In this chapter, we focus on the purchasing function. In the following chapter, we address the EHS function.

Purchasing professionals have been flooded with new purchasing programs over the last two decades. These include:

- Just in time (JIT)
- JIT II
- Supplier consolidation
- Value-added purchasing
- Leveraging
- Integrated supply chains
- Supply chain management
- Outsourcing
- Insourcing
- Relationship marketing
- Partnerships
- Systems contracting

Some work wonderfully, others are slow and difficult to implement, yet others are simply passing fads. In this chapter, we explain how chemical management, and, particularly, Shared Savings, relates to trends in the purchasing field. We explore supplier partnerships—what they are and when they make sense. Finally, we look at how chemical management can change and benefit the purchasing function.

CHEMICAL MANAGEMENT
AND PURCHASING TRENDS

I came to this company two years ago specifically to do integrated supply, take a large number of suppliers and leverage the business, get additional value-added services in addition to price savings, and come up with a better total cost of ownership deal. Now, I can do all that and more with chemical management.

—*A plant purchasing manager*

Sorting out the barrage of new purchasing programs can leave purchasing managers weary and frustrated. Riggs and Robbins (1998), in their recent work on supply management, make the following observations:

[M]any procurement philosophies, ranging from "Just get it here!" to intense leveraging for price, from harsh adversarial relationships to fuzzy, feel-good partnerships, have left procurement professionals unsure of their roles. . . .What seems to be lacking in this endeavor is a way to consistently use supply resources to improve overall business performance.

In the companies we studied, Shared Savings chemical management can help promote and integrate these new supply strategies. But whether Shared Savings fits comfortably with these initiatives depends on the purchasing philosophy of the company. In its simplest form, the distinction comes down to this: Which of the following statements best describes the chemical purchasing philosophy at your company?

A. We know what we want in chemical products and services; we simply need to find the supplier who can reliably provide these at the lowest cost.

B. Chemicals are not our core business. We want a supplier who has the expertise and ability to increase chemical performance and reduce total chemical costs.

We encountered many successful companies whose purchasing philosophy is best described by statement A. They invested heavily in developing their own internal chemical management capability and see chemical management as part of their core business. However, they may begin to pay the price for this approach as their investment in chemical management drains resources from activities more central to their core business. They may find themselves slipping behind competitors who have tapped chemical supplier expertise.

Chemical management fits best with companies following philosophy B. In fact, chemical management complements many purchasing initiatives *if* the initiatives focus on supplier involvement rather than supplier response to the demands of the buyer. Below, we examine several purchasing initiatives and how they relate to chemical management.

Consolidation and Leveraging

Supplier consolidation is an integral component of chemical management. To improve performance, a supplier must be involved with a significant range of chemicals used in the plant, particularly those chemicals that can affect the performance of others. For example, certain coolants are compatible only with certain cleaners designed to remove them. When the supplier has a broad scope of chemical responsibility, it can improve overall chemical performance.

Chemical management also employs volume leveraging, but not just for price concessions. In chemical management, leveraging is used to provide one supplier with sufficient volume to compensate for the significant investment in personnel and equipment that is often required in chemical management programs. In other words, volume leveraging offers the opportunity for both sides to enter a high-performing, synergistic relationship that would not be economically or operationally feasible at lower volumes. (In some cases, volume and historical margins may be sufficiently high to allow price concessions *in addition* to covering administrative costs.)

Outsourcing and Insourcing

Chemical management may be considered a form of insourcing, as opposed to the more traditional practice of outsourcing. The difference is *where* the work is being performed. Insourcing can be defined as "transferring work done *in-house* to a supplier who performs the work *in-house* (i.e., the supplier moves on-site)" (MacNabb 1997). In chemical management, chemical suppliers typically take over many of the responsibilities and activities previously assigned to plant personnel. Outsourcing, on the other hand, transfers to an outside supplier work previously performed by in-house personnel.

However, there is an important difference between chemical management and common insourcing. In most insourcing programs, the plant separates itself from the insourced activities. Those activities become the sole responsibility of supplier personnel. Chemical management is better described as an *integration* of supplier personnel into the plant's basic operations. Both plant and supplier personnel are involved in optimizing chemical processes.

Integrated Supply Chain Management

Integrated supply chain management involves coordinating the movement of goods and materials across the companies in a supply chain. It typically includes supplier consolidation, expanded services, and a closer relationship between buyer and supplier.

Chemical management can fit well into integrated supply chain management initiatives. For example, some of the auto companies involve selected chemical management suppliers in product and production design to avoid chemical-related problems in production. In some instances, the supplier reengineers a chemical to meet a customer's unique production or application specifications. Similarly, chemical management complements many related purchasing initiatives, including systems contracting, blanket contracting, and value-added supply.

SUPPLIER PARTNERSHIPS

What we wanted was performance, the value-added contribution that [a supplier] could bring to this plant.
—A plant purchasing manager

Supplier partnership is one of the most common terms used (and misused) in describing today's supply strategy initiatives, yet the term *partnership* is differently interpreted by different buyers. A clear definition of this word is needed to understand chemical management and its role in the broader procurement strategy of a company. Below, we offer our view of the fundamental characteristics of a partnership and the circumstances under which it can benefit a buyer. We also present four lessons for successful partnerships derived from Shared Savings chemical management experience.

What Is a Supplier Partnership?

Through our study of chemical supply partnerships, we identified one fundamental criterion for distinguishing supply partnerships from traditional supply relationships: *the creation of value*. The traditional supplier relationship is essentially a zero-sum game in which buyer and supplier battle over price and volume. Gains for one side create losses for the other. These are sometimes termed *transactional relationships* because the focus is on the transaction: exchanging money for product. In a transactional relationship, both sides believe that value in the relationship is created through the transaction itself.

In contrast, a partnership is based on the belief that value is created in the *performance* of the product or service. Unlike a transactional relationship, unlimited financial rewards can be achieved for both parties as they work together to increase value for the ultimate consumer.

A partnership can begin with a subtle yet fundamental shift in the financial structure of the relationship. For example, a buyer may pay a premium on a chemical because the supplier's technical expertise proved valuable in reducing production problems. Both the chemical user and chemical supplier benefit from the expansion of the relationship, which is no longer driven by price and volume alone. Of course, this is only the beginning of a partnership. Yet it reflects the fundamental purpose of a partnership: creating business value. As we have shown, that requires performance, not just the delivery of chemicals.

When Are Supplier Partnerships Appropriate?

Our research findings suggest that these two criteria best indicate when supplier partnerships are most appropriate (see Fig. 11-1):

1. The ratio of impact to purchase price is relatively high.
2. The supplied product or service is not within the buyer's core business.

Impact, the first criterion, includes effects on quality, capability, and total cost of ownership (TCO; see chapter 7). We used an iceberg to represent the TCO for chemicals, but a similar iceberg can be created for any supplied product or service. In those cases where the purchase price (the tip of the iceberg) is large relative to the rest of the iceberg, the impact-to-price ratio tends to be smaller, on average. On the other hand, when

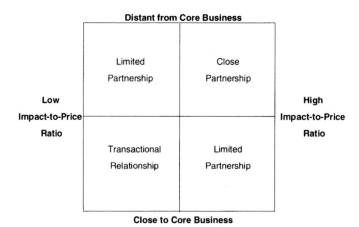

Figure 11-1. Partnership recommendations under alternative supply conditions.

the purchase price is small relative to the rest of the iceberg, the impact-to-price ratio is larger, on average. For most chemicals, the purchase price is often a small fraction of the TCO, resulting in a large impact-to-price ratio. This large ratio seems to be typical of indirect purchases for most manufacturers. Riggs and Robbins (1998) suggest that purchase price averages only about 25–50% of the TCO for manufacturing and assembly industries. Similar assessments can be performed for quality and capability impacts.

Core business refers to a company's market and its key competitive competencies in that market. Managers in the plants we studied routinely used phrases such as "building quality engines" and "producing the cars that customers want" to describe their core business. In these companies, chemicals were described as the lifeblood of the manufacturing operation, but everyone agreed that their companies were not in the chemical business.

These two criteria are critical because they help determine when managers should look outside their company to create greater value for their customers. In Fig. 11-1, we presented these criteria as a 2x2 matrix. When purchased products or services have a low impact-to-price ratio, purchase price should be the dominant determinant of value, and it makes sense to place greater emphasis on getting the lowest possible price. An example for some companies might be common office supplies. In other cases, especially when the impact-to-price ratio is *very* large, the proper use and management of the chemical should become the dominant determinants of value. When the ratio is large, changes in purchase price are almost insignificant in the overall creation of value for the ultimate consumer and the profitability of the chemical buyer.

When the impact-to-price ratio is relatively high, chemical application and management expertise is essential. The chemical user has two options: to develop and maintain this chemical expertise internally or to seek assistance from sources outside the company, such as the supplier. It is becoming increasingly evident that a key to competitive success is *organizational learning*. Companies are continually pressured to invest their limited resources in learning to maximize return on investment. This concept applies to all aspects of the organization, from chemicals to computers. Monetary and nonmonetary returns on investment in organizational learning are typically highest when a manager invests in the company's core business because a company's core competencies are the basis for its competitive strength. Investing resources in developing knowledge and expertise in areas remote to the company's core business diverts them from areas that provide a higher return on investment. Instead, it is more cost-effective to purchase the knowledge and expertise from a company whose core business includes that knowledge and expertise.

Consider, for example, a chemical supply company where regulatory compliance represents a significant proportion of the company's total cost of chemical ownership. Chemical manufacturers, as part of their core business, must track regulations related to their chemicals and develop compliance programs for their manufacturing facilities. In the partnerships we studied, the chemical suppliers provided guidance and expertise to develop compliance programs in the automotive manufacturing plants that used their chemicals. This expertise produced significant savings in terms of labor and time for the plants. Even greater benefits resulted from applying the supplier's chemical expertise to monitoring and improving chemical performance in the plant. Remember, this expertise is not simply provided as a service, as the chemical supplier's bottom line improved through improved chemical performance.

When evaluated together, these criteria suggest that partnerships are best suited for products or services with a high impact-to-price ratio that are not an integral part of the buyer's core business. Our research suggests that this include not only most chemicals but many other indirect purchases as well, including information technology. At the other extreme of the 2 × 2 matrix are products and services with a low impact-to-price ratio that actually are part of a company's core business. In this situation, a transactional relationship is appropriate, as the purchase price is the focus.

When the impact-to-price ratio is high and the product or service is within the buyer's core business, more limited forms of partnership may be appropriate. For example, the chemical management partnerships common to the automotive industry proved less popular in the electroplating industry in part because electroplaters commonly define their core business as "electro-chemistry" (Bierma and Waterstraat 1997b). Investment in their own chemical expertise produces a high rate of return; thus, electroplaters have less need for the expertise of their chemical suppliers. However, many still seek their suppliers' advice regarding issues of new technologies, water treatment, waste management, and even basic process chemistry. Many electroplaters noted that they have difficulty keeping up with regulations and new developments in chemistry. We found an interest in more limited partnership arrangements, such as gainsharing, to encourage suppliers to develop and share beneficial innovations.

In cases where the impact-to-price ratio is low but the product or service falls outside the buyer's core business, a limited partnership is probably the best relationship as well. The purchase price of the chemical is the major cost component and thus a transactional relationship best serves the needs of the chemical buyer as well as supplier.

Four Lessons for Partnerships

Four key lessons for successful partnerships arise from our research of Shared Savings chemical management programs. These lessons are important not only for partnerships in the supply of chemicals but also are applicable to the supply of all types of products and services.

1. *Purchase price is the tip of the iceberg*—Focus on TCO and performance leverage.
2. *Pay for performance, not volume*—Value is added when the product or service performs, not when it is received. Link supplier revenue to performance.
3. *Focus on core business*—Learning is the key to competitive strength. Focus on core business and partner with those who have the complementary core expertise you need.
4. *Reward continuous improvement*—Drive continuous improvement through natural consequences of improved value and profit, not threat.

CHANGES IN THE ROLE OF PURCHASING

The concern for buying leverage is now secondary to the interest in improving the fortunes of both buyer and seller.
—C.C. POIRIER AND W.F. HOUSER,
Business Partnering for Continuous Improvement

Purchasing plays a key—and, in some companies, a lead—role in all of the chemical management programs we studied. However, the responsibilities and the associated skills required to fulfill those responsibilities are different than those required by the traditional chemical supply approach.

Transactional Supply—Purchasing Skills and Responsibilities

The traditional assignment for most chemical purchasing programs is to find the supplier with the lowest price who can meet the quality and delivery specifications of engineering and production departments respectively. Evaluation of purchasing programs and personnel may even hinge on their success in controlling prices. Traditional skills concentrate on understanding and searching the supply markets, managing the bidding process, and negotiating the best terms.

However, this analysis does not begin to capture the complications that arise in chemical purchasing. These include:

- Buying hundreds or even thousands of different chemicals. Some chemicals, such as coolants, may have over a dozen different brands specified by the users.
- Processing rush orders resulting from poor inventory monitoring programs.
- Coping with large numbers of small-volume orders.
- Tracking thousands of purchases made directly by chemical users as part of a program to reduce the number of small-volume orders handled by the purchasing department.
- Dealing with the side effects of direct vendor-user relationships, such as brand specifications and chemical trial samples.

As a result, most of Purchasing personnel time is spent handling paperwork or putting out fires. Little time is available to improve the strategic value of the purchasing function or the chemicals that it buys.

Chemical Management—Purchasing Skills and Responsibilities

The purchasing function changes dramatically under chemical management. After the initial awarding of the chemical management contract, many companies only rebid the contract once, if at all. Suppliers typically handle chemical ordering and payment; purchasing departments may issue only one chemical purchase order each year.

The purchasing function shifts from that of tactical supply to a strategic development role. Often taking the lead in managing chemical management programs, Purchasing acts as the interface between the strategic interests of the company and the Tier 1 chemical supplier, including such activities as:

- Working with or leading plant chemical management teams to determine chemical performance and management needs.
- Searching the market to find suppliers with the best technology and expertise to meet those needs at a competitive price.
- Developing and monitoring of key metrics for total chemical costs and chemical performance.
- Benchmarking total chemical cost and performance metrics against other chemical users inside and outside of the industry.
- Performing should-cost analyses on selected supplier activities to assure competitive pricing.

These changes require Purchasing personnel to develop both market and technology expertise. The way Purchasing interacts with other departments changes, as does the supplier selection process—from bidding for the lowest price to securing the best supplier at a competitive price. Less ordering and more supply-stream management occurs. As a result, the purchasing function becomes a key and visible contributor to business value and competitive strength.

IMPLICATIONS FOR THE MANAGER

In many ways, Shared Savings chemical management is a natural fit with many purchasing initiatives. Consider the following questions. Is your purchasing department attempting to:

- receive greater value-added services from your suppliers?
- reduce the time and paperwork involved in ordering supplies?
- reduce inventory by better coordinating supply with production?
- stem the continuous rise in chemical costs?

Shared Savings programs contribute directly to these goals. However, these programs can conflict with many current purchasing practices. For example, in Shared Savings programs:

- The plant no longer buys chemicals. Instead, it buys chemical expertise and services.

- Chemical purchasing decisions are made by a chemical management team; actual chemical purchasing is done by the supplier.
- The purchasing department produces just one purchase order per month, or perhaps just one per year.
- Supplier payment is unrelated to the volume of chemical supplied.
- The contract might be bid only once. After that, it might simply be renegotiated.
- The contract price might increase in order to help overall costs go down.

The involvement and support of Purchasing management is essential to the success of any Shared Savings program. However, a number of difficult changes may be required. At the plants we studied, Purchasing personnel raved about the benefits of Shared Savings even though the transition had not always been an easy one!

12

EHS and Chemical Management

This program provides much more help in maintaining control than I have ever had.
—DAN UHLE, *Environmental Engineer, Ford Chicago Assembly Plant*

Environment, Health, and Safety (EHS) personnel play a critical role in achieving a successful chemical management program. In return, chemical management programs can dramatically improve EHS performance. To understand the role of EHS personnel in chemical management, it is useful to understand and appreciate EHS's unique perspective on production operations—*waste*. In this chapter, we begin with a review of waste and the benefits of waste minimization. We then examine the EHS benefits that can result from chemical management and the important role that EHS personnel play in making chemical management successful. Finally, we explore how chemical management can change the EHS function in a company.

WASTE—EHS'S UNIQUE PERSPECTIVE

Everyone thinks of waste as an environmental issue. That's a natural mistake.
Waste is . . . the biggest opportunity North American manufacturers
have ever had to increase their profits.
—*Charles Rooney, Orr and Boss*

Waste is a loss. It is company resources going down the drain. It produces a negative return on investment. After all, if you think about it, waste is composed of material that a company purchased. Labor and capital resources are applied to it in the production process. Finally, the company treats it so it won't be harmful, and then throws it away! Or, worse yet, they pay someone else to throw it away!

We have seen that production operations are the source of quality and the source of costs. They are also a source of waste. Fig 12-1 illustrates a production operation similar to that in Fig. 6-2. It is composed of inputs, a transformation activity, and outputs. However, the product is not the only output of an operation. If inputs do not become product, they become waste. (We define *waste* as any output that has a value less than the cost incurred to produce it. Thus, all waste represents a net operating loss to the company.)

There are two basic sources of waste from an operation. One is composed of inputs intended to become part of the product. If an operation is not 100% efficient at converting such inputs into product, some of them become waste. For example, overspray in a painting operation does not become part of the product and must either be disposed of or reclaimed. Sheet metal, plastic, cardboard and other materials may not be fully used in a process, leaving scrap. The product itself may become waste if the quality is poor or it is damaged in an operation.

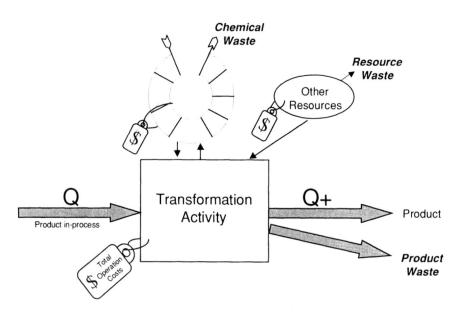

Figure 12-1. Operations generate waste.

Waste can also be generated in ancillary operations that support production. For example, the process of managing coolant for a machining operation can generate waste at several points in the chemical life cycle. Spills occur during receipt and storage of the coolant. Inventory goes out of date. Pumps supplying coolant to machining equipment leak. Once the coolant has served its purpose in an operation, it becomes waste at the end of the chemical life cycle.

THE COST OF WASTE AND THE VALUE OF WASTE MINIMIZATION

Nobody buys material just to throw it away. All waste contains material and labor.
—CHARLES ROONEY, *Orr and Boss*

The cost of waste is much more than the cost of disposal (see Table 12-1). It includes the purchase value of the material in the waste and handling, storage, and treatment costs. The wastes may contribute to health and safety hazards in the workplace as well as EHS compliance costs. Clearly, a company is both financially and operationally better off if it can find ways to generate less waste from its operations. Generating less waste—waste

Table 12-1.
Common Components of Waste Cost

1. *Material costs*—As stated in the opening quotation of this section, "nobody buys material just to throw it away." The purchase price of the material can often be one of the largest waste costs.
2. *Treatment costs*—Many wastes must be treated before being released into the environment.
3. *Disposal costs*—The cost of proper disposal can vary from simple discharge to hazardous waste haulage service.
4. *Other life cycle costs*—Chemicals and other resources have in-plant life cycles, from receiving to treatment and disposal. All these costs are reduced when less waste is generated.
5. *Lost value-added*—Product waste may pass through many value-adding operations before becoming waste. For example, sheet metal may be cut, drilled, formed and painted before becoming damaged in a packaging operation. The cost of this added value is included in waste.
6. *Compliance costs*—Certain wastes may make a company subject to environmental, health, and safety (EHS) regulations. This can generate costs ranging from additional paperwork to buying new equipment and building new structures.

minimization—typically requires improving operation inputs or transformation activities, or both, to achieve maximum chemical performance with minimal waste. This can have a significant positive effect on business value.

The total cost of waste can be enormous. In a study of waste costs in various industries, Charles Rooney (1993) found that *waste costs often equal or exceed direct labor costs*. Waste minimization reduces these costs by preventing the waste. End-of-the-pipe control as a means of accomplishing environmental performance objectives only adds to the total cost of waste. A waste-minimization strategy not only meets (or exceeds) environmental objectives but *reduces* the total cost of waste. As we show below, it can enhance other aspects of business value.

WASTE MINIMIZATION
ENHANCES BUSINESS VALUE

[W]aste, or pollution, is a key indicator of unnecessary cost, and,
most important, inattentive routine management.
—F. MCINERNEY AND S. WHITE, *The Total Quality Corporation*

Beginning in the mid-1980s, more companies began to recognize that the generation of waste represented inefficiencies or defects in the production process. As part of the larger quality revolution sweeping through American business at the time, these companies found that it was far cheaper to improve the process and generate less waste than it was to treat the waste once it was generated. As a result, many companies achieved levels of environmental performance well beyond those required for compliance while making significant contributions to the bottom line. Federal and state governments promoted this effort through numerous pollution-prevention initiatives. In addition to providing significant cost savings, companies found that waste minimization offered the benefits of increasing quality and capability.

Quality

In some cases, waste minimization activities result in enhanced product attributes. They can also enhance the reputation of both product and company. For example:

Alcan Rolled Products Company (Lagoe 1996)—The company, which reclaims scrap aluminum, used large amounts of chlorine gas to clean, or flux, molten aluminum. Much of the chlorine was released as hydrogen chloride emissions, producing downwind damage to vegetation and buildings. Growing use of chlorine in the process raised additional concerns about future regulatory compliance. To reduce chlorine emissions, the company investigated a new fluxing process that cleans the metal as it leaves the melting furnace. The new technology reduced chlorine emissions by 95%. Moreover, it provided a cleaner aluminum product, increasing levels of customer satisfaction. The dramatic reduction in chlorine use decreased the likelihood of a chlorine leak or employee injury, enhancing the reliability of production rates.

Majestic Metals, Inc. (Johnson and Burnap 1995)—This small sheet-metal fabricating and coating plant undertook a wide array of small but significant waste minimization projects beginning in the early 1990s. These projects resulted in significant improvements in environmental performance, attracting the interest of the media and public officials. The company signed three new industrial customers as a result of its new environmental reputation.

Chrysler Corporation (see case study in chapter 17)—Faced with tightening regulations on emissions from painting operations, the Chrysler assembly plant in Belvidere, Illinois, worked with its paint supplier to develop a powder coat anti-chip painting process. The new process not only reduced emissions dramatically, it significantly improved the durability of the paint finish. Chrysler also worked with the paint supplier to develop a method for correcting paint defects that not only required less solvent but resulted in a higher-quality finish.

Locomotive repair facility—A Midwest locomotive repair facility fitted half of its parts washers with membrane filtration devices to continuously clean the washing fluid, reducing the frequency with which fluid was dumped and replenished. This not only reduced fluid discharges but also provided a consistently cleaner fluid to the parts washers, producing a more stable cleaning process and a cleaner product.

Consumers often shop for environmentally preferable, or "green," products. At many of the plants with chemical management supply programs, the suppliers have been instrumental in working with manufacturers to improve their environmental image, including elimination of ozone-depleting substances as well as chemicals on the Toxic Release Inventory (TRI) list. These changes improve marketability.

Capability

Waste minimization can have a profound effect on a company's ability to generate value in the future. Most importantly, waste minimization enhances a company's adaptability. Many waste-minimization efforts simplify processes and reduce cycle times. In the Alcan example cited above, for example, the new flux process reduced two steps in the production process and shortened cycle time by nearly 20% (Lagoe, 1996). As we show in the case histories later in this book, Shared Savings chemical management programs enhance adaptability through the close working relationship established with the chemical supplier. Chemical products, processes, and procedures can be changed quickly because of the improved communication and cooperation between the chemical supplier and the manufacturer.

CHEMICAL MANAGEMENT BENEFITS FROM EHS LEADERSHIP

Improved environmental performance has been a basic focus area
of this program from the very beginning.
—Dr. P.N. Mishra, *Chemicals Management Program Coordinator,*
General Motors Corporation

EHS must play a central role in chemical management supply programs. EHS staff possess both the technical background and the cross-functional contacts needed to see the big picture and coordinate chemical policies, procedures, and applications in a plant. Their responsibilities involve them with all of the plant's chemical processes and the personnel who use them. In addition, they are often familiar with the entire life cycle of chemicals in the plant, from receiving to treatment and disposal.

In some plants, the EHS manager coordinates the entire chemical management program. Working with Purchasing, EHS oversees everything from supplier selection to the daily operations of the plant chemical management team. In other plants, Purchasing or another department may coordinate the program, but EHS plays a lead technical role. In each of the five case histories contained in this book, EHS staff either directed the programs or played a key role on the plant's chemical management committee.

General Motors (GM) assigns a major role to environmental staff in plant chemical management programs. In many plants, the environmental manager is the program coordinator. At the Janesville Truck and Bus plant, featured in chapter 16, Mike Merrick, a senior environmental engineer, coordinates the program. Chemical management at GM originated in the corporate environmental division and the environmental staff continue to lead the effort today. GM's primary emphasis is on *improving processes that use chemicals*, not simply improving the logistics of chemicals. EHS is in the best position in the GM facilities to fulfill this total management strategy.

Dr. P.N. Mishra of GM, one of the creators of chemical management for the automotive industry explains, "As far as General Motors is concerned, the program is technology based. It is a technology program. . . . [T]he intent is to step up into higher levels of technology continuously." Further, "process improvement and process optimization . . . are the keys to this whole program. The dollars come as a result." (Mishra 1997b)

EHS BENEFITS FROM CHEMICAL MANAGEMENT

> We have seen so many benefits from the program—
> better information, better inventory control, better
> management of chemical use, emergency planning support. . . .
> —LINDA LITTLE, *Environmental Engineer, GM's Janesville Truck and Bus Plant*

Chemical management programs address many of the causes of EHS problems. Reducing the number and volume of chemicals provides substantial EHS benefits, but a Shared Savings chemical management program produces a broader range of EHS benefits as well. While, individually, any one of these benefits might be achievable without a chemical management program, it would be difficult to accomplish such a consistent and dramatic improvement in environmental performance without the total involvement of the chemical supplier enjoyed under Shared Savings.

Below, we present many of the common EHS challenges found in a typical plant: controlling process wastes; reducing other chemical wastes; compliance with recordkeeping, reporting, and training regulations; and employee health and safety.

Controlling Process Wastes

Controlling process wastes with end-of-pipe technology is expensive and difficult. Equipment upsets and downtime are regular occurrences as the volumes and composition of wastes change unpredictably.

A chemical management program improves this situation because cross-functional coordination of activities and departments is a key objective. Supplier personnel are on site specifically to help bring chemical processes under control and reduce chemical waste. In fact, improving processes and reducing waste increases supplier profit in Shared Savings chemical management programs! Cross-functional coordination of chemical use and the associated EHS improvements were the benefits most frequently cited by EHS personnel in our interviews.

An excellent example of cross-functional chemical management effectiveness is solvent applications. Used in plants for many purposes by many processes and personnel, solvents can be a major contributor to total volatile organic compound (VOC) emissions. Controlling solvent VOCs is difficult and time consuming. However, suppliers with expertise in solvent applications can work one-on-one with each user in the plant to find more efficient application methods and materials. The following comment from Dan Uhle at Ford's Chicago Assembly Plant, is typical:

> We felt we were making every attempt possible to reduce our level of solvents, but we're approaching it even more aggressively with the [chemical management] contract. [Our supplier] is helping us to reduce VOCs with much more vigor than we could have on our own.

As a result, Ford was able to reduce VOCs by 57% in eighteen months. The GM truck plant at Fort Wayne, Indiana, had a similar experience. The chemical management program was able to cut purge solvent usage in half over a two-year period.

Wastewater treatment systems are easier to operate under a Shared Savings chemical management program because the supplier working with the wastewater treatment operators is the same supplier working with the production personnel, maintenance personnel, and others who generate the wastewater. In fact, it is in the supplier's financial interest to minimize wastewater treatment problems because it owns the treatment chemicals.

Reducing Other Chemical Wastes

Chemical waste comes not only from production processes but also from chemical products that simply never get used. For example:

- Inventory wastage (off-spec or outdated chemicals)
- Spillage
- Discontinued usage
- Overpurchase due to supplier volume-discount policies
- Incorrect chemical purchases
- Supplier samples

These problems can be reduced dramatically with a Shared Savings chemical management program. Chemicals belong to the supplier until they are used in the plant. Suppliers actively work to minimize the in-plant chemical inventory, significantly reducing the likelihood of spillage, spoilage, incorrect volumes, and the use of wrong chemicals. Unused supplier samples, which had caused disposal problems in the facilities we studied, remain the responsibility of the chemical supplier, eliminating potential disposal problems.

Bob Conrad, environmental specialist at Chrysler, explains his company's experience:

> We don't own the chemicals; [our supplier] is responsible for them. If there is a problem with a chemical, they take it back or do what's necessary to correct the problem. We don't own the chemicals until they have worked successfully on our product. The point is that you pay only for a quality finished product. You pay your supplier for products that are actually salable.

Complying with Recordkeeping, Reporting, and Training Regulations

An array of regulations requires specific chemical information to be collected, reported, and, communicated to workers. Two of the more demanding compliance activities are informing workers about safe use of chemicals (including access to material safety data sheets [MSDSs]) and completing the annual chemical reporting documents (TRI). Training and reporting standards require accurate, current data regarding the identity, toxicity, and volume of chemicals used in the plant, as well as the disposition of those chemicals.

Few plants have the extensive chemical tracking systems needed to routinely collect these data. In many plants, it is difficult to identify the total volume of chemicals used in a given year, or even the number of chemicals used. The responsibility for chemical purchasing and recordkeeping is distributed across several departments or operating units and most companies do not have the resources to implement plantwide chemical tracking systems. Records are kept largely for financial and accounting reasons, not for purposes of chemical management.

Chemical tracking is frequently one of the key activities included in a chemical management program. Many suppliers have developed tracking software and expertise and consider this one of their most important competitive advantages. While the primary goal of chemical tracking is usually production and process improvement, it offers many EHS benefits. TRI reporting is greatly simplified, with data on chemical receipts, inventory, usage, and wastes reported monthly to EHS by the supplier. The supplier coordinates MSDS documentation and may also provide computerized storage and access to MSDSs as well as employee education and training.

Again, the experience of Dan Uhle at Ford's Chicago plant is typical:

> Chemical tracking has improved environmental reporting substantially. That used to be the hardest part in doing our Form Rs—coming up with good chemical usage data. . . . Take solvents, for example. I had to look at what we bought and get records from the suppliers of all of the different materials that had VOCs in them. I had to make the assumption that the inventory at the beginning of the year was the same as at the end of the year. I didn't know exactly how much was scrapped and actually went out as waste paint solvent, where we had some recovery, or how much was actually emitted. So I had to make assumptions about all of that to the best of my ability using engineering judgment. . . . Now PPG keeps daily records of what they use and the VOCs emitted, and they not only do it by product but they do it by process.

Health and Safety

Reducing the number and volume of chemicals in the plant improves health and safety conditions in the workplace. Having fewer chemicals means that chemical handling and spill response procedures are more manageable. Having less volume reduces the likelihood and potential magnitude of a chemical accident as well.

In addition, many suppliers include health and safety compliance assurance as part of their services. Experience from their own chemical manufacturing facilities or those of Tier 2 suppliers led them to develop successful health and safety programs, which they can offer to their customers. Moreover, supplier personnel rely heavily on the cooperation of plant employees for proper chemical use and handling. They regularly look for opportunities to improve the employees' knowledge and understanding of chemical application. Suppliers offer frequent education and training opportunities for plant personnel.

An Example of EHS Benefits

The EHS benefits of chemical management are too numerous to list or discuss in their entirety. However, as an illustration of what is possible, we provide a list of chemical management examples from GM that resulted in reduced number, volume, or toxicity of chemicals (Table 12-2), and a list of other EHS benefits experienced by the GM truck assembly plant in Fort Wayne, Indiana (Table 12-3).

Table 12-2.
Examples of Chemical Management Program Activities
Resulting in Reduced Number, Volume, or Toxicity of Chemicals
at General Motors Plants (Knoblock 1998)

- Use totes instead of drums.
- Use steam condensate for non-contact cooling.
- Reclaim solvent at source.
- Use oil blend to meet viscosity targets.
- Replace current chemicals with more efficient chemicals.
- Use hot water for purge instead of solvents.
- Share solvent containers.
- Purchase concentrates instead of diluted product.

- Optimize concentration of chemicals in process tanks.
- Eliminate chemical usage.
- Eliminate chemical inventories.
- Increase cycles on cooling towers.
- Automate chemical feed systems.
- Optimize chemical feed and usage.
- Consolidate product.
- Replace solvents with environmentally friendly chemicals.
- Use electrostatic oil cleaners in machines to reduce the number of oil changes needed.

Table 12-3.
Other EHS Benefits at the GM–Fort Wayne Assembly Plant
(Knoblock 1998)

- Additional resources for technical, clerical, and hazardous materials support
- Centralized chemical information management, including inventory tracking
- More information with greater accuracy
- Assistance in waste management activities
- Additional qualified personnel at plant for DOT, RCRA, HAZMAT, etc.

CHANGES IN THE ROLE OF EHS

It's like an ombudsman—we coordinate the overall program
while the supplier manages the details.

—RUDY BERNATH, *Plant Chemist, Navistar International Engine Plant*

The traditional role of EHS is to assure compliance with applicable EHS regulations. Managing waste is a concern only after the fact—*after* the waste is generated in the production process. Referring to our diagram of overall business operations (Fig. 7-3), the EHS manager protects the company's license to operate. That is, EHS prevents the company from being unplugged by regulatory authorities. Unfortunately, this is an expensive strategy, incurring significant costs just to maintain the firm's license to operate. These costs drain shareholder value; EHS programs are generally regarded as a necessary evil for the company.

Under Shared Savings, the biggest change most EHS personnel notice is having more time to proactively address EHS problems. While the plant retains responsibility for EHS performance and regulatory compliance, the supplier supports related efforts. EHS departments provide direction, but suppliers perform many of the daily activities, such as testing and recordkeeping. As the number and volume of chemicals in the plant declines, EHS personnel devote more time to improving environmental performance.

This means a change in the activities of EHS personnel. Examples include:

- A move from compliance activities to compliance audits
- Reduced paperwork and searching for information
- Reduced usage and monitoring of pollution-control equipment

- Reduced usage of personal protective equipment and associated activities (training, respirator fit testing, etc.)
- Reduced time spent finding and interpreting regulations, or developing compliance plans
- Increased time spent working with Production, Engineering, and other departments that use chemicals
- Increased time spent monitoring total chemical costs and performance
- Increased time spent coordinating chemical management activities

These changes require the EHS manager to function less like a technical specialist and more like a business manager. The new role requires broader technical expertise and more business management skills. Technical expertise is needed in the area of chemical applications—how chemicals perform in the workplace. Business skills, such as the use of financial and cost accounting tools, are needed to monitor costs and performance and to quantify the business value added by EHS and the chemical management program.

IMPLICATIONS FOR THE MANAGER

Waste literally bleeds a company of its resources. Waste adds no value and can cost a significant amount to dispose of, yet it is often overlooked by almost every department in a plant—except EHS. This puts EHS personnel in a unique position to apply their technical skills to improve processes, reducing both waste and cost.

Shared Savings chemical management offers an excellent opportunity to integrate EHS with overall business goals and demonstrate the link between EHS and business success. Shared Savings can also produce changes in the EHS function. In particular, it requires a switch from reliance on end-of-the-pipe controls to a focus on process optimization; it requires a switch from compliance to profit as justification for capital investment; and it requires reliance on supplier personnel to perform many day-to-day chemical management activities. These may not be easy changes, but the benefits can be well worth the effort.

13

Concerns and Misconceptions about Shared Savings

Implementing Shared Savings chemical supply relationships in a company requires a significant investment of time and effort, and the initial risks can seem high. It requires important changes for many organizational divisions and their personnel. Old policies, procedures, and habits need to be reviewed and revised. Even though the benefits can be dramatic, it is not surprising that most personnel are hesitant to adopt Shared Savings. In our research, we found they often have one or more of the following concerns or misconceptions.

"I'VE ALREADY GOT MY OWN CHEMICAL MANAGEMENT PROGRAM."

Many companies have developed their own sophisticated chemical management programs that allow them to successfully achieve some degree of chemical control. It is not surprising that these firms are reluctant to adopt a new, untested program. But adopting Shared Savings doesn't mean abandoning an existing chemical management program. *It requires defining a new role for suppliers that promotes and rewards supplier performance.* Any existing program must be revised to tap supplier initiative and resources by allowing suppliers to reap a portion of the financial benefits of their innovations and efforts.

Some companies may not wish to have an increased level of supplier involvement in their manufacturing operations. Shared Savings is not an appropriate strategy for them. However, these companies must continue to expend resources to develop their own chemical expertise or risk falling behind their competition. When their competitors are able to tap the initiative and resources of their suppliers, it is difficult for a company using traditional chemical supply to remain competitive. Plants have cut millions of dollars in operating costs from their budgets, increased product quality, and improved production processes by utilizing their suppliers' expertise once they adopted a Shared Savings program.

"MY SUPPLIER ALREADY PROVIDES CHEMICAL MANAGEMENT SERVICES."

Suppliers offer many chemical management programs with an array of services under many names. A supplier may monitor chemical quality, manage inventory, or dispose of waste chemicals. These can be valuable services. However, *caveat emptor* applies when seeking a vendor to provide Shared Savings services. Individual services or program names are not always a good indicator of the actual type of program a vendor is promoting. Suppliers may create new products or services to provide what *appears to be* a Shared Savings program, but careful analysis of the program is required to assure that it *is* a true Shared Savings program. When examining existing or proposed supplier programs, refer to the list of program characteristics in Fig. 8-1 and consider the following questions:

- What is being purchased—chemicals, chemical services, or chemical performance? In Shared Savings programs, suppliers are typically paid for chemical performance and retain ownership of the chemical until it is used in a process.
- What type of volume incentive does the financial arrangement create for the supplier? Will the supplier make more money if chemical volume increases or decreases? In Shared Savings, supplier profit increases as chemical volume decreases.
- Can the supplier profit from cost-saving ideas it brings to the company? In a Shared Savings relationship, the suppliers has the opportunity to share directly in the financial benefits created for the customer. This type of relationship allows the chemical user to leverage

the supplier's expertise and resources—even in areas not directly related to chemicals.

- What services, if any, are provided by the supplier (logistic, environment, health, and safety (EHS)/compliance, or application)? Are the services provided on-site or off-site? It is usually most advantageous to have on-site service and support in all three areas.

- Are services based on a thorough analysis of the company's chemical needs? Are they consistent with the core competencies of the supplier? Ideally, services address clearly defined needs of the chemical user and match the core competence of the chemical supplier. For example, water treatment chemical suppliers may provide machining fluids, but it is not their core business. If a company's primary needs are machining fluids, it is better to find a supplier whose core competency is in machining fluids. To obtain the greatest value from the relationship, the supplier's expertise should be consistent with the user's primary needs.

These simple questions can assist the manager with objectively evaluating a Shared Savings program offered by a chemical supplier.

"SUPPLIER COMPETITION WILL KEEP MY CHEMICAL COSTS DOWN."

Many chemical suppliers are offended by the notion that they are driven to increase chemical volume rather than fulfill their customer's chemical needs. Suppliers often argue that competition in the chemical industry provides sufficient incentive for them to save their customers money through reduced chemical use or other services.

Nevertheless, logic and experience suggest a fundamental difference in the behavior of suppliers when their compensation is linked to chemical volume rather than chemical performance. While suppliers may compromise their short-term financial gain in order to keep an account, this is not the same as applying their company's resources to help continuously reduce a customer's chemical use.

Perhaps the best way to understand the difference in supplier motivation is to consider the supplier's primary concern under different chemical supply strategies—the primary questions that drive chemical supply staff every day:

- If the chemical supply program is *transactional*, the primary question is "How can I sell more product?"
- If the chemical supply program is *service-oriented*, the primary question is "How can I keep this account?"
- Under a Shared Savings program, we believe the primary question is "How can I help the customer make a better product with less chemical at a lower cost?"

At this point, it is tempting to think we exaggerate and to conclude that no supplier could actually be as altruistic as the Shared Savings question above suggests. But we propose that it is the transaction-oriented and service-oriented suppliers who are expected to be altruistic. After all, is it reasonable to expect suppliers to ignore their own financial interests and do what is in their customers' financial interests? Even if they do, how much profit are suppliers willing to sacrifice to make their customers financially successful? Under Shared Savings, however, suppliers are expected (and paid) to pursue *their own* financial interests—to make all the money they can make! Under Shared Savings, suppliers are paid for *value, not volume*. Suppliers and their customers have the same financial interests. As one chemical sales representative explained, "All our arrows are heading in the same direction." Increased consumer value leads directly to increased profit for both the supplier and the customer.

"I CAN'T TRUST MY SUPPLIER."

Shared Savings poses a new dilemma for the chemical user: "If my supplier makes more money by reducing chemical costs, what is to stop the supplier from doing this by cutting corners, creating a bigger chemical problem for me in the future?" For example, a supplier of boiler water treatment chemicals might substitute lower-quality chemicals in order to reduce costs, resulting in increased corrosion of boiler pipes that must be replaced at the expense of the chemical user.

There is no simple solution to this dilemma, but our research suggests that a successful Shared Savings program requires at least two conditions. First, both parties must enter into the relationship with a *long-term* perspective. Both parties must focus on the long-term benefits of the relationship rather than the short-term benefits gained through cutting corners. The long-term perspective applies to both the supplier and the user.

Second, the manufacturer must clearly understand its own systems well enough to develop appropriate quality and performance specifications. For example, a portion of the supplier's compensation may be linked to the annual amount of boiler pipe corrosion. While chemical quality or performance specifications can never address every detail, a few key performance parameters can focus both parties on long-term results.

For example, one auto plant implemented a unit price program for its paint detackification chemical supplier without implementing a full Shared Savings program. In other words, there was no attempt to build a long-term mutual commitment nor was the system studied well enough to develop adequate quality and performance specifications. The supplier did reduce chemical volumes, but detackification sludge increased dramatically, increasing overall disposal costs for the automaker. Supplier revenue had not been sufficiently linked to total chemical performance. The supplier responded to the new financial incentives and the automaker suffered the consequences.

"I DON'T WANT TO GIVE UP CONTROL."

Control is a good thing for a manufacturer. However, it is important to clarify what *control* means. In our interviews, when people used the word *control,* they meant one of two things: (1) the ability to make and implement better decisions, or (2) the ability to make and implement decisions without supporting data.

Invariably, the people we interviewed at plants with Shared Savings programs insisted they had greater control over performance and quality than ever before. The increased control was a direct result of improved operating and performance data as well as the chemical expertise and resources at their disposal.

In contrast, many of the individuals who expressed fear over losing control under Shared Savings were concerned not about performance but changes to their own authority and prerogatives. Individuals who are currently able to make and implement decisions without having to demonstrate *with data* that each is the best decision for the company find Shared Savings a direct threat to this type of control. Such people often strenuously oppose Shared Savings. As one EHS manager said regarding the importance of data in a Shared Savings program decision-making process, "In God we trust; all others must bring data."

"I'LL MISS MY PERFORMANCE TARGETS."

The purpose of a Shared Savings supply relationship is to improve the long-term financial performance of the company. For some managers, this may result in missing short-term financial performance targets. However, this reflects a problem with the structure and timelines of the financial performance targets, not the Shared Savings supply relationship.

As we showed in chapter 2 and 3, optimizing the annual budget of one department can seriously compromise long-term benefits for the entire company. Most managers' budgets do not reflect the total costs of chemicals and do not include costs incurred by other departments.

In addition, Shared Savings programs may increase expenses in the short term while the program is in a ramp-up period. New storage and handling equipment, improved information technology, or employee education and training may require increased short-term investments to provide long-term gains. While a Shared Savings program may cause the short-term budget targets for some departments to be missed, it improves the long-term benefits for the company as a whole.

"IT THREATENS JOBS."

Shared Savings appears to threaten jobs in two ways. First, if the supplier assumes responsibility for in-plant chemical management activities, the chemical user might have less need for its own in-plant personnel. Second, because Shared Savings focuses on eliminating waste, the chemical user might have less need for personnel currently involved in managing those wastes.

Our research findings support the conclusion that Shared Savings frees up employee time. The issue is what the chemical user does with the additional personnel time? Some managers do use it as an opportunity to reduce staff; however, practice has demonstrated that Shared Savings initiatives don't work as headcount reduction programs. One major automaker found that plant managers who perceive Shared Savings programs as headcount reduction opportunities experience immediate implementation problems.

In companies where Shared Savings has been successfully implemented, managers redirected employee resources to operations that are closer to the company's core business, producing *higher value per employee hour*. In the companies we studied, personnel reported that Shared Savings improved job quality—sometimes in dramatic ways. Fritz Benton, on-site chemical manager for BetzDearborn at the GM Truck and Bus plant, tells of just one example:

> We were going into another plant with a [Shared Savings] contract that included the wastewater treatment plant. The chief operator was getting ready to retire at the time of the contract. However, after we came in and provided advice and helped make a lot of process changes, he said his job had improved so much that he decided to stay another two years!

Shared Savings creates new roles for many personnel. They become more integrated with the core business of the company, working with managers in Purchasing, Production, Materials Handling, Maintenance, and other areas that interface with the supplier through the Shared Savings program. Here is what Dan Uhle, plant environmental engineer at the Ford's Chicago Assembly Plant, had to say:

> Before this program and others like it, there was no light at the end of the tunnel for my responsibilities. . . . Now that has all changed. Certainly one of biggest benefits with this program is that we have been able to get the management help that is necessary for continuous training, continual improvement in waste reduction, and even in improving quality. We've been able to do this without raising costs, and in some instances the costs have gone down.

"IT WILL CREATE A LIABILITY NIGHTMARE."

Many companies considering Shared Savings express serious concerns about liability. Who is responsible if there is an accidental chemical release into the environment? Or a chemical fire? What if a chemical user's employee spills the chemical supplier's chemicals? When chemical supplier personnel operate inside the chemical user's plant, who is responsible if they cause damage or a chemical release? The possible legal scenarios are endless.

We posed the question of liability to supplier and company managers with years of Shared Savings experience. The people we interviewed, both chemical users and chemical suppliers, usually responded to this question with a look of puzzlement. They could not understand why liability would be more of a concern in a Shared Savings relationship than in any other supply relationship.

The comments of Jerry Mittlestaedt, environmental manager at Navistar's engine plant in Melrose Park, Illinois, are typical of the answers we received. "Why are they so afraid of liability?" he asked. "Isn't that part of doing business?" Bob Hendershott, on-site supplier representative at the Navistar plant, added, "You have to show responsibility. If something is happening that you don't think is right, document it. Write a letter. It's the same in a sales situation. If you are not being responsible and an accident happens, you probably deserve to be liable for that."

Though each company had certain provisions for liability in their Shared Savings contract, the people we interviewed agreed that whether the relationship is traditional or a Shared Savings relationship, both companies must take responsibility for their personnel, products, and decisions. As Ted Camer, on-site supplier representative at Ford's Chicago Assembly Plant, put it "I guess we understand the liability instinctively, so it's not something we spend much time thinking about."

"IT WILL BREACH CONFIDENTIALITY."

It is reasonable for companies to be concerned about a chemical supplier being intimately involved with the operation of proprietary production processes; there is a potential for proprietary ideas to be shared with competitors who are serviced by the same supplier. Though confidentiality protection clauses can be a component of any Shared Savings contract, there is no guarantee that trade secrets will not be passed inadvertently through supplier personnel located at the competitors' plants.

The worst-case scenario is a company whose sole competitive advantage is based on a confidential process technology and whose chemical supplier also does business with the company's primary competitors. This could present a significant barrier to implementing Shared Savings. However, it is in these companies' best interest to develop means of protecting proprietary technology while still taking advantage of the benefits of Shared Savings. Creative strategies are needed to find a Shared Savings approach that will work under this circumstance. This may require more specific contractual language or limiting supplier access to areas in the plant that proprietary operations. At a minimum, Shared Savings relationships may still be possible for ancillary chemical applications, such as water treatment chemicals or maintenance paints.

IMPLICATIONS FOR THE MANAGER

We have spoken about Shared Savings at many conferences and business meetings. The subject always generates considerable response from the audience—but the response consists almost solely of statements about why Shared Savings would never work in their plants. Few people ask questions about how the plants we studied manage to make Shared Savings work despite facing the same problems. From these many encounters, we developed our list of common concerns and misconceptions.

Once in a while, one person in the audience says, in essence, "The benefits of Shared Savings seem simply too great to ignore. How can I make this work in my plant?" It is for these people that we have prepared the remaining chapters of this book. In part 3, we share examples from companies that have years of experience with Shared Savings. In part 4, we explore the elements of Shared Savings in greater detail and discuss how a plant *can* make Shared Savings work.

PART 3

SHARED SAVINGS IN ACTION

CHAPTER

14

Chemical Management Case Histories

One of the best ways to understand chemical management, especially Shared Savings chemical management, is to see it in action. We developed case histories of Shared Savings chemical management programs at five manufacturing and assembly plants in the midwestern United States. These are presented in chapters 15–19.

No one knows exactly how many plants currently use Shared Savings chemical management supply strategies. Most companies consider their chemical management programs a key competitive advantage and are reluctant to share their knowledge and experience openly for fear of losing the competitive advantage. These companies have made significant investments of time and effort to develop their programs; therefore, few are willing to divulge technical or economic details.

It is clear that Shared Savings has a strong foothold in many companies and continues to grow in acceptance. Companies with experience in Shared Savings chemical management continue to implement it in more of their plants. It has been used in the North American automotive industry for over a decade and is being adopted in the electronics, airlines, and even forest products industries. It is apparently also being used by selected companies in the steel, appliance, aircraft, and food and beverage industry (Mishra 1998).

Suppliers with strong chemical management capabilities continue to report growth in their programs. Castrol Industrial North America, for example, reports that more than 50% of the staff in their metalworking fluids division are working under chemical management agreements.

Before presenting the five Shared Savings case histories in the following chapter, we summarize, below, chemical management programs from a variety of industries.

AUTOMOTIVE INDUSTRY

GM-Canada—Oshawa, Canada
(Reid 1997)

The General Motors of Canada complex in Oshawa assembles both cars and trucks, with an annual capacity of over 750,000 vehicles. A Shared Savings chemical management program was implemented in 1996 covering paint detackification, paint stripping and masking, purge solvents, oils and lubricants, parts washers, and water treatment chemistry. The Tier 1 supplier provides on-site management, an on-site laboratory to control chemical processes, inventory management, and chemical tracking and reporting.

The program produced immediate benefits. In 1996 alone, chemical inventory was reduced 17%. Chemical costs declined 11% in 1996 compared to 1995, with a further reduction of 5% in 1997. GM also enjoyed the immediate elimination of costs associated with chemical purchasing and inventory management. Of the many improvements in environmental performance, the net volume of purge solvent at the truck plant was cut in half during 1996 and 1997 (see Fig. 14-1). This reduction was accomplished largely through a cooperative effort between supplier and plant personnel to improve the efficiency of solvent use.

GM—Fort Wayne, Indiana
(Knoblock 1998)

General Motor's truck assembly plant in Fort Wayne, Indiana, assembles light duty pickup trucks in a three-million-square-foot assembly facility. The Shared Savings chemical management program footprint includes paint shop chemicals (other than paint), water treatment, oils and lubricants, compressed gases, general maintenance chemicals, and janitorial

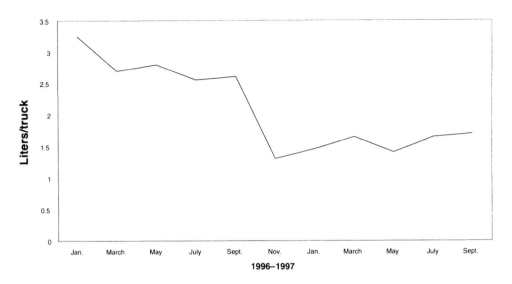

Figure 14-1. Purge solvent net usage, GM–Oshawa Truck Plant, 1996–1997 (Reid 1997).

chemicals. The supplier, Texo Chemicals Management, provides a full range of on-site chemical management services.

The program cut purge solvent usage from 1.2 gallons to 0.77 gallons per truck, reducing volatile organic compound (VOC) emissions by over 700,000 pounds. Process changes at the plant prior to chemical management required increased paint booth cleaning, resulting in dramatically higher labor and chemical costs. Process changes made through the chemical management program cut cleaning frequency in half, reduced cleaning labor hours by 4,000 hours per year, saved an additional 50% in material costs, and produced a 93,000-pound reduction in VOC emissions. An effort to improve overall paint booth management improved paint transfer efficiency 5% (saving 0.1 gallons of paint per vehicle) and cut detackification chemicals and paint sludge by 5%.

Navistar International—Springfield, Ohio
(Wiseman and Knight 1996)

Navistar International's plant in Springfield, Ohio, assembles 80% the company's International line of trucks. The plant implemented a chemical management program in conjunction with its paint shop chemical supplier, Ashland Chemical. Efforts to improve paint shop procedures and chemical use reduced solvent usage by 138,000 gallons per year and paint usage by 65,000 gallons per year, saving more than $3,000,000 annually on a volume-adjusted basis.

Land Rover—Solihull, United Kingdom
(*Machinery Market* 1992)

The Land Rover plant in Solihull in the United Kingdom entered into a Shared Savings chemical management program with Castrol Ltd. Recognizing that coolant costs represent less than 1% of operating costs but affect more than 80% of operating costs, the plant sought to improve coolant management as a means of reducing operating costs and increasing performance. Under the contract, Castrol provides on-site management of all the metalworking fluids.

In the first sixteen months of the program, the number of coolant types used in the plant was cut from twenty-six to nine. Coolant disposal declined from 230,000 to 40,000 liters per month. Machine downtime was reduced dramatically and the number of machines taken out of service for sump cleaning each week was cut by 50%. With chemical management, tool life increased more than 10% as well.

ELECTRONICS AND AEROSPACE INDUSTRIES

Nortel—Ottawa, Canada
(Broe 1997; ENDS Report 1997; Votta 1998; Votta et al. 1998)

Nortel Semiconductors, a unit of Northern Telecom Ltd., Canada, became interested in chemical management as other electronics industry leaders, including Motorola, began adopting such programs. Working with its supplier, Olin Chemical, Nortel began a chemical management program in 1994 at its semiconductor facility in Ottawa, Canada.

The program began as a limited chemical management arrangement. Olin provided chemical procurement, on-site testing, distribution, and other services in return for a management fee. Nortel's benefits include reduced usage of acids, caustics, and organic coatings. The company also experienced reduced waste generation and improved container management.

Two years later, however, Nortel joined the Chemical Strategies Partnership (CSP) to assist in upgrading the chemical management program to a Shared Savings program. CSP, a non-profit organization based in San Francisco, was created to promote chemical management, particularly in the electronics industry. CSP is funded by combined grants from the Pew Charitable Trusts and the Heinz Endowments. Nortel featured the Shared Savings approach to chemical supply in its 1996 environmental progress report.

Lockheed Martin—Syracuse, New York
(Daniels 1998)

Lockheed Martin initiated an unusual chemical management program with Radian International, which provides chemical management services to facilities in Lockheed's electronics sector. Radian is not a chemical supplier but instead a manager of chemicals provided by Tier 2 suppliers. The primary management activities involve chemical logistics and consolidating chemicals and suppliers. Radian provides comprehensive inventory services and has eliminated minimum-buy problems through multifacility volume purchases. It has improved the chemical tracking and material safety data sheet (MSDS) management systems. Lockheed Martin has enjoyed over $1 million in savings from the program, largely through reductions in procurement and hazardous waste expenses.

AMP, Inc., and Raytheon Electronics Systems—
Harrisburg, Pennsylvania
(Krampe 1998; Moshure 1998)

AMP, Inc., and Raytheon Electronics Systems are in the process of developing chemical management programs as part of the CSP.

AIRLINES

Delta Air Lines—Atlanta
(Craig 1998)

Delta Air Lines' Atlanta maintenance facility is responsible for the maintenance and overhaul of a fleet of over 550 aircraft. Historically, the facility purchased 1,472 different chemicals annually, totaling more than $18 million per year. Delta established a logistics-oriented chemical management program with a regional firm (Interface LLC Chemical Management), which acts as a Tier 1 supplier. The supplier purchases and retains ownership of all chemicals and provides off-site warehousing and just-in-time delivery services to the users.

Delta has realized a significant improvement in chemical logistics, including better chemical tracking at lower cost, a shorter procurement cycle, $200,000 annual savings in insurance increases, $250,000 reduction in inventory spoilage, and a 25–30% reduction in overall material usage.

FOREST PRODUCTS

Weyerhaeuser—Tacoma, Washington
(Loren 1996)

Weyerhaeuser, one of the largest forest products companies in the world, took a different approach to implementing chemical management supply programs. It began with an extensive four-year internal chemical management development effort that involved input from facilities across the United States and Canada. This led to the recognition that better chemical supply relationships with vendors would enhance the chemical management effort:

> [F]acilities can now see where they might have been penny wise and pound foolish in the past. . . . [O]ne full-service solvent supplier that charges a higher per-unit cost but deals with hazardous materials paperwork may suddenly become more attractive than their lower-priced partial-service competitor. Consequently, Weyerhaeuser is finding that many of its facilities are moving to single-source suppliers.

FIVE CASE HISTORIES

The authors developed the case histories in chapters 15 through 19 through a series of interviews and site visits conducted between 1995 and 1998. Both chemical user and chemical supplier personnel were interviewed, ranging from union employees to senior management at the corporate level. Together, these five plants have over thirty-five years of experience in Shared Savings chemical management.

The following are summaries of the five plants and their Shared Savings programs. All cases are published with the consent of both chemical supplier and chemical user companies.

Navistar Engine Plant, Melrose Park, Illinois

Began first Shared Savings program: 1987

Name of current program: Chemical Management Services (CMS)

Supplier: Castrol Industrial North America

Chemical footprint: Machining coolants, cleaners, and associated additives

Financial terms:
- Fixed monthly fee based on historical chemical usage
- Staffing fee for full-time in-plant support personnel
- Formula for sharing large financial gains or losses, including rebates to Navistar for large reductions in chemical usage.

Performance expectations: Quality and performance specifications, including rust prevention

Supplier services:
- Chemical acquisition
- Inventory and distribution management
- Container management
- Quality assurance and maintenance oversight
- Testing and lab analyses
- Product and process engineering development
- EHS studies and training
- Process and waste problem solving

Benefits:
- Coolant usage reduced by over 50%; coolant waste haulage reduced by over 90%
- Rework on blocks and heads reduced by 93%
- Rebates for reduced chemical usage up to $10,000 per year
- Reduced production downtime
- Reduced wastewater loading
- Improved inventory control and reduced inventory costs
- Improved health and safety protection
- Easier compliance reporting
- Illinois Governor's Pollution Prevention Award

Ford Chicago Assembly Plant, Chicago, Illinois

Began first Shared Savings program: 1988

Name of current program: Total Fluids Management, Total Solvents Management

Supplier: PPG (Chemfil division)

Chemical footprint: Most chemicals in the plant, with the exception of paints, sealers and lubricants

Financial terms:
- Fixed fee per vehicle (unit pricing) based on historical chemical usage and production
- Fixed annual fee for selected chemicals that cannot be linked to production volume.

Performance expectations: Expected annual reductions in the fee per vehicle as well as a productivity reduction—a negotiated level of savings to be realized by Ford

Supplier services:
- Chemical acquisition from Tier 2 suppliers
- Inventory and distribution management
- Container management
- Quality assurance and maintenance oversight
- Testing and lab analyses
- Product and process engineering development
- VOC emission reduction training
- EHS studies and training
- Process and waste problem solving

Benefits:
- VOC emissions reduced by 57% in eighteen months
- Wastewater sludge generation reduced by 27%, saving over $50,000 per year
- Steady or declining chemical costs
- Improved product finish quality; reduced rework
- Improved inventory control; reduced inventory costs
- Improved health and safety protection
- Easier compliance reporting
- Illinois Governor's Pollution Prevention Award application

Chrysler Belvidere Assembly Plant, Belvidere, Illinois

Began first Shared Savings program: 1989

Name of current program: Pay-As-Painted

Supplier: PPG

Chemical footprint: All chemicals and systems related to cleaning, treating, and coating auto bodies

Financial terms:
- Fixed fee per vehicle painted to quality specifications (unit pricing) based on historical chemical usage and production

Performance expectations: Quality and performance specifications drive the agreement. PPG's fee is based on producing a specified paint finish.

Supplier services:
- Chemical acquisition
- Inventory and distribution management
- Container management
- Quality assurance and maintenance oversight
- Testing and lab analyses
- Product and process engineering development
- EHS studies and training
- Process and waste problem solving

Benefits:
- Over $1 million in savings from the first year of Pay-As-Painted
- Dramatic reductions in VOC emissions and other wastes
- Improved product quality; reduced rework
- Improved inventory control; reduced inventory costs
- Improved health and safety protection
- Easier compliance reporting
- Four Illinois Governor's Pollution Prevention Awards

General Motors Truck and Bus Plant, Janesville, Wisconsin

Began first Shared Savings program: 1992

Name of current program: Chemicals Management Program (CMP)

Supplier: BetzDearborn

Chemical footprint: Water treatment chemicals (powerhouse, cooling towers, wastewater treatment, air houses), paint detackification and booth maintenance, lubricants, maintenance paints, commodity chemicals, purge solvents

Financial terms:
- Fixed fee per vehicle (unit pricing) based on historical chemical usage and production
- Management fees for selected services

Performance Expectations:
- Unit prices to be steady or declining.
- Targets for overall plant savings, including annual savings, equal to 5% of the value of the contract

Supplier Services:
- Acquisition and inventory control
- Monitoring and coordination of chemical usage
- Chemical performance research and improvement
- Chemical and regulatory testing lab
- Ongoing reporting and communication
- Product and process engineering development
- EHS compliance and training
- Continuous waste minimization

Benefits:
- Over $1 million in savings
- 8% decrease in chemical costs with significantly expanded services
- Improved inventory control; reduced inventory costs; product consolidation
- Training and other programs to improve health and safety protection
- Chemical tracking for easier compliance reporting
- Reduced VOC emissions and sludge disposal
- Reduced downtime and labor cost for sludge clean-out
- Many other improvements that reduce labor overtime, improve process efficiency, improve product quality, and reduce rework

GM's Electro-Motive Division, LaGrange, Illinois

Began first Shared Savings program: 1994

Name of current program: Chemicals Management Program (CMP)

Supplier: D.A. Stuart Company

Chemical footprint: Machining fluids (coolants), cleaners, oils, water treatment, and miscellaneous small-volume chemicals

Financial terms:
- Fixed monthly fee based on historical chemical usage and production
- Management fees for selected services

Performance expectations: Annual fee reductions of 6% for three years and 3% for the next five years

Supplier services:
- Acquisition and inventory control
- Monitoring and coordination of chemical usage
- Chemical performance research and improvement
- Ongoing reporting and communication
- Product and process engineering development
- EHS compliance and training
- Continuous waste minimization
- Filter management

Benefits:
- Greater than 40% reduction in chemical costs
- Greater than 50% decrease in coolant usage and coolant waste together with coolant performance increase
- Elimination of biocide additions of central coolant systems
- Elimination of sodium nitrite from washers; reduction in cleaner usage and waste
- Improved inventory control; reduced inventory costs; product consolidation
- Training and other programs to improve health and safety protection
- Chemical tracking for easier compliance reporting
- Reduced VOC emissions
- Many other improvements that reduce labor overtime, improve process efficiency, improve product quality, and reduce rework

15

Navistar Engine Plant

Shared Savings Case History
Navistar International
Engine Plant
Melrose Park, Illinois
and
Castrol Industrial North America, Inc.

Bob Hendershott had recently designed and implemented a new tracking system for a cleaning solution. Hendershott worked at Navistar International's engine plant, where parts washers are used at many points in the manufacturing process. The new system tracked cleaning solution usage by individual washer, not just department, and on a much more frequent basis than in the past. It wasn't long before he identified a problem. One machine was using far more cleaner than the others. A follow-up analysis by Hendershott found that the washer had a faulty filling mechanism, spilling cleaner into the overflow each morning when the machine was turned on. Because the faulty mechanism was inside the washer, no one had ever observed the problem. Without Bob's new tracking system, the leak might have gone undetected for years.

Management was thrilled! Fixing the filling mechanism dramatically reduced cleaner usage as well as cleaner discharge to the wastewater treatment plant, saving money and reducing waste. Hendershott received some well-deserved recognition for his work. But not just from Navistar. In fact, he did not even work for Navistar. He worked for the company that supplied the cleaning solution to Navistar, and the supplier, his employer, was just as thrilled!

SUMMARY

The Navistar International engine plant in Melrose Park, Illinois, has a decade-long Shared Savings relationship with Castrol Industrial North America. The fixed-fee contract, which covers coolants, cleaners, and associated additives, has produced dramatic results. The financial incentive for Castrol to improve chemical use efficiency at the plant has resulted in a reduction in coolant usage of more than 50% and a reduction in coolant waste of more than 90%.

But the benefits are not limited to chemical volume. Navistar experiences less production downtime and improved product quality. Potential production, health, and environmental problems are identified and resolved more quickly, before they become significant. Compliance reporting is much easier, given the chemical tracking data provided by Castrol. Overall, the opportunity for each company to focus on its core business has produced superior performance and profitability across the board.

THE NAVISTAR ENGINE PLANT

The Navistar Engine Plant in Melrose Park, Illinois, employs 1,200 people in a 1.5-million-square-foot manufacturing facility. Originally built in 1941 to make aircraft engines for the B-24 bomber, the plant now produces about two hundred diesel truck and bus engines per day. In 1986, the Navistar International company was born from the dissolution of the International Harvester company after it sold its farm equipment division.

The plant uses a wide array of chemicals, from oils, coolants, and cleaners in manufacturing to water treatment chemicals in auxiliary operations such as cooling towers, wastewater treatment, and boilers. Of these chemicals, coolants and cleaners predominate due to the number of machining operations in the plant.

CASTROL INDUSTRIAL NORTH AMERICA

Castrol Industrial North America, Inc. is one of several metalworking fluid suppliers who provide advanced chemical management services to their customers. Castrol has participated in Shared Savings chemical supply programs since the mid 1980s. Its program can be tailored to meet the specific needs of almost any manufacturing customer but generally involves a wide array of chemical services through on-site management and a fixed-cost structure for billing.

THE NAVISTAR-CASTROL RELATIONSHIP

It saves us tons of money and improves the environment.
This whole thing is just good business.
—JERRY MITTLESTAEDT, *Environmental Control Manager*

The Contract

As with any major manufacturing facility, Navistar uses a wide array of chemicals. The Castrol Shared Savings contract focuses on two high-volume groups of chemicals: coolants and cleaners (and related additives). Other chemical suppliers provide most of the chemicals outside these two groups. Navistar has not yet established Shared Savings relationships with these other suppliers.

The program uses a fixed monthly fee that was established based on historical chemical use data (see Table 15-1). This arrangement provides a significant incentive for Castrol to improve chemical use efficiency and cut waste.

A portion of the savings from improved efficiency is returned directly to Navistar through a rebate program, which has provided as much as $10,000 per year in cash benefits.

The People

Castrol provides a full-time on-site manager, Bob Hendershott, who is assisted by a full-time on-site technician to support an array of services. Though the Castrol representatives oversee logistic, compliance, and application services, Navistar's United Auto Workers (UAW) members perform the hands-on work. This has promoted a productive, cooperative relationship between Castrol and the union. Bob Hendershott emphasizes that the Shared Savings contract "was not implemented as a headcount reduction program at Navistar."

The Castrol on-site personnel work directly with Bob Monroe, Navistar's head of the Machining Business Team Unit (BTU), who reports to the plant manager. Hendershott provides monthly progress reports to the

Table 15-1.
The CMS Contract at Navistar

1. **Chemical footprint**—coolants, cleaners (excluding solvent and aerosol), and associated additives
2. **Financial relationship**
   - Fixed monthly fee for coolants and additives
   - Fixed monthly fee for cleaners and additives
   - Additional staffing fees
3. **Risk/Reward**—formula for sharing unusually large financial gains or losses. This has produced annual rebates to Navistar for large reductions in coolant usage. It also covers significant shifts in chemical costs.
4. **Responsibilities**—lists the responsibilities of Navistar and Castrol. Navistar UAW employees provide most of the hands-on work, including chemical changes and additions. Castrol activities include:
   - Purchasing
   - Inventory and distribution management
   - Container management
   - Quality assurance and maintenance oversight
   - Testing and lab analyses
   - Product and process engineering development
   - EHS studies and training
   - Process and waste problem solving
5. **Liabilities**—sets out division of liabilities for chemical management and performance, including rust prevention. Liability from chemical losses prior to unloading from the truck remain with Castrol. Losses after unloading are allocated based on who was responsible.
6. **Performance requirements**—expected chemical volume and cost reductions as well as specific projects to be completed by both Castrol and Navistar
7. **Customer feedback system**—procedures followed by Castrol to assess the level of satisfaction of its Navistar personnel customers

Machining BTU, including chemical usage and savings, but also has a close, day-to-day working relationship with supervisors and other Machining BTU personnel on the plant floor. He also works with other departments involved with plant chemicals, including Purchasing, Environmental Control, and Health and Safety. He serves on several chemical-related committees and chairs the Coolant Committee for the plant.

Though the contract specifies the services to be provided by Castrol personnel on a day-to-day basis, Bob Hendershott assumes responsibilities beyond the contract. His workday is defined more by the needs of Navistar's production process than by the details of the contract. As he says "I get to the point where I really start to feel like I'm a Navistar employee. I am looking out for both sides." With activities ranging from anticipating future regulatory problems to providing coverage for the plant chemist while he's on vacation, Hendershott routinely exceeds the scope of the contract to meet the needs of Navistar. Though Castrol receives no fees directly for these additional services, they contribute to the strong customer loyalty the company has earned at Navistar. As with other successful Shared Savings relationships, the strong personal working relationships established between Castrol and Navistar personnel are most remarkable. After years of working through problems together, both sides continue to sing each other's praises. This explains why the ten-year relationship between Castrol and Navistar is stronger today than ever.

EVOLUTION OF THE RELATIONSHIP

Navistar is in business to make engines, the best engines they can.
Castrol is in the business of coolants, rust preventatives,
cleaners, and that's what we do 100% of the time.
—BOB HENDERSHOTT, *Castrol On-site Manager*

Great business relationships often emerge from humble beginnings—and stressful circumstances. This is clearly evident at Navistar.

Getting Started—The Hardest Part

The 1980s was a period of reawakening for many American industries, and the automotive industry was no exception. It experienced one of its worst slumps in history and took a severe economic beating from a slowdown in the economy and from foreign competition. On top of this, the downturn in the farm economy forced International Harvester Company to sell off its farm equipment lines and reform itself as Navistar International. In the mid-1980s, Navistar plants were downsizing and searching for efficiency improvements. At the Melrose Park engine plant, downsizing meant focusing on core business, reducing costs, and improving operational controls. The company was seeking new ways of accomplishing more with less. For Rudy Bernath, the plant chemist, this meant finding a way to perform the numerous responsibilities of the burdened chemical management staff with fewer people.

In 1985, Bernath was approached by a Castrol representative with a novel concept: make Castrol the sole supplier of coolants for the plant. In return, Castrol would accept a flat monthly fee for the fluids at a rate below Navistar's current monthly coolant bill. In addition, Castrol would perform many of the routine monitoring tasks that Bernath's staff were struggling to complete, such as testing and maintaining the coolant systems throughout the plant. The fixed-fee arrangement ensured that both Castrol and Navistar would benefit from improved chemical management and reduced chemical use in the plant .

The Navistar engine plant was to be one of Castrol's first experiences with what they call their Castrol + Plus^{TM} program, or, more generally, Chemical Management Service (CMS). Today, CMS is Castrol's fastest-growing business segment.

Overcoming Resistance

Though CMS provided Navistar with a variety of benefits, the initial champions of the program, the plant chemists, saw it as an opportunity to refocus their limited resources on activities aligned closer to the company's core business—production, quality control, and health and safety. In addition, Navistar stood to benefit from the stable chemical costs and assistance in reducing environmental discharges. Rudy Bernath and his chemical staff worked long and hard to gain the support and approval of key Navistar decision makers to implement the new contract. In the beginning, it was rough going. CMS was a hard sell to the various Navistar departments and divisions involved in setting up the new contractual arrangement.

Some of the sources of Navistar's concern were:

- *Manufacturing units*—At that time, Navistar used many companies to supply manufacturing chemicals. Castrol was just one of several coolant suppliers. The manufacturing units in the plant were skeptical about replacing suppliers they had used for years. The thought of relying solely on a single supplier made the manufacturing units very uncomfortable.
- *Purchasing*—Purchasing personnel were initially concerned about setting up an open purchase order. Even though the monthly fee was fixed, a purchase order that did not specify quantity and unit price was hard to swallow. It simply was not consistent with past practice and experience.

- *Maintenance*—Maintenance personnel had responsibility for maintaining machining equipment and fluids. Their concern was that mistakes by Castrol would increase their workload.
- *Legal*—Liability was the primary concern of the legal department. The CMS contract represented a new type of business relationship. Castrol representatives would be performing tasks that previously were performed by Navistar personnel. It was unclear how responsibilities and the associated liabilities would be assigned. The means of terminating the relationship also needed to be specified.

It took over a year, but eventually all the key Navistar personnel, including the plant manager, took the risk and approved the contract. "It took strong support from plant management," notes Brian Nordman, the first on-site Castrol representative, now a Castrol site manager at Delphi Saginaw Steering Systems. "It was like a marriage, and Castrol had to earn everyone's trust over time."

The First Contract

To evaluate CMS on a limited basis, Navistar gave Castrol one of the five central coolant systems in the Machining BTU. The 70,000-gallon system supporting the engine block machining process was the system giving Navistar the most problems. The fee was fixed at the previous year's average monthly expense, with a small discount. Castrol assumed responsibility for monitoring the use and condition of all Castrol chemicals in the system. A part-time field representative from Castrol provided these in-house services to Navistar.

Navistar realized immediate benefits from the new relationship. The chemical staff were able to devote more time to quality control and process improvement, with noticeable results. With the routine chemical monitoring provided by Castrol, machining fluids were maintained consistently within operating specifications, resulting in less machine downtime and fewer machining defects. Incidents of chemical injuries (e.g., dermatitis) also declined. The Castrol personnel were experienced with adjusting the fluids to meet performance specifications while controlling chemical concentrations to reduce personnel hazards. Inventory control improved and costs declined. Wastewater discharges declined dramatically as coolant waste was reduced.

An additional benefit was the reduction of scrapped and reworked engine components. Through active management of process chemicals, the performance of the coolants and cleaners improved, machining defects dropped, and rusting incidents declined. Rework on engine blocks and heads declined 93%. These factors contributed to reducing production costs by $1.50 per engine. Within the first few years of the contract, skepticism at Navistar began to disappear, and both sides developed a strong, close working relationship.

Full Time

After almost three years, a full-time Castrol CMS site manager was hired. The CMS contract was expanded to cover all five central coolant systems. The on-site manager's responsibilities were expanded to include chemical purchasing, inventory control, and distribution as well as an active role in production improvement. Purchasing and chemical management costs for Navistar continued to decline.

The on-site manager also began to take a more active role in several of the quality teams that Navistar had organized, particularly the Coolant Team and the Safety Team. These teams initially had trouble getting off the ground and functioned below their potential. The on-site manager, however, served as a focal point and a resource for these groups and helped them become proactive and productive in addressing chemical management issues in the plant.

It was also in this phase of the relationship that Castrol began to significantly reduce chemical usage. Castrol and Navistar personnel worked together to identify processes and applications where Castrol products would solve fluid use and management problems in the plant. Many coolant applications were switched over from soluble oils to synthetics, extending coolant life and reducing fluid carry-out on the parts. On Castrol's recommendation Navistar purchased a portable coolant repolishing unit that removed contaminants and tramp oils, thus extending the life of the fluid, improving fluid performance and reducing waste.

The CMS manager worked on resolving incompatibilities between coolants and cleaners. Previously, chemical incompatibility problems had resulted in spotting on the product, reducing quality. The changes recommended by the on-site manager eliminated the spotting problems and improved product quality. Castrol provided all the analysis and testing resources required in its own laboratories, reducing research and development (R&D) expenses for Navistar.

THE BENEFITS

"It's saved us a lot of money on raw materials and waste disposal," is the reflection from Jerry Mittlestaedt when asked about the Shared Savings relationship. "I'm amazed at what a great program this is." Mittlestaedt is the manager of environmental control at the Navistar plant and an enthusiastic advocate of this innovative chemical supplier relationship.

"You can look at the bottom line," says Castrol's Hendershott. "You can look at production over time and the reduction in chemical usage. Production is steady or climbing and chemical usage is going down; waste is going down." Fig. 15-1 illustrates this for coolant usage at the Melrose Park plant. Though production has increased, coolant usage has been cut by over 50% and coolant waste haulage has been cut by over 90%.

These kinds of benefits are the most dramatic and easily measured—but they are not the only ones (see Table 15-2). Coolants and cleaners are critical to the proper operation of machinery, the health and safety of the workers, and the quality of the product. "Not properly managing our coolants and cleaners could *stop* production," comments Mittlestaedt. "If we don't run our assembly line at two hundred engines in eight hours, that's a million dollars in lost production. If there is a problem with a part like a crankshaft . . . you could shut this assembly line down. I can't tell you how important this [Castrol relationship] is."

Navistar has also benefited from Castrol's expertise in coolants and cleaners. As Hendershott summarized it, "you've got one rep here, but in addition to that, you've got all the Castrol people that are behind that rep who are here basically every day. The other day, health concerns were raised about a coolant that had been in the system for quite some time. I called our toxicologist to assist us with the problem."

Sanjay Patel, a Navistar environmental engineer, added to that point. "If we have a problem, such as a metal creating a problem with a particular coolant, you can bring that back to Castrol and they will adjust the product to solve the problem. I don't think we would have the capability to do it on our own. These people [Castrol] are professional. They are always researching ways to make this better." To that, Hendershott responds, "Navistar is in the business of making engines, the best engines they can. Castrol

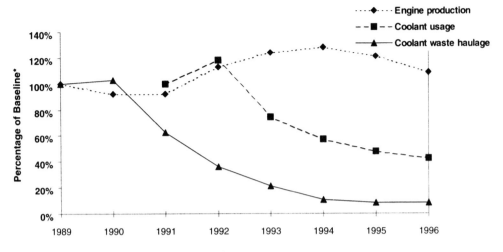

Figure 15-1. Engine production, coolant usage, and coolant waste haulage, Navistar engine plant, 1989–1996.

is in the business of coolants, rust preventatives, cleaners, and that's what we do 100% of the time."

Because chemicals are Castrol's core business and chemical usage is directly linked to cost, Castrol tracks chemicals more efficiently and effectively than Navistar can. Castrol personnel track chemical usage to individual machines and sumps rather than to units or divisions. They constantly monitor machine usage and chemical inventory levels. This has provided valuable

Table 15-2.
Selected benefits for Navistar

- Coolant usage reduced by over 50%; coolant waste haulage reduced by over 90%
- Rebates for reduced chemical usage up to $10,000 per year
- Reduced production downtime
- Improved product quality, reducing engine block and head rework 93%
- Reduced wastewater loading
- Improved inventory control and reduced inventory costs
- Improved health and safety protection
- Easier compliance reporting
- Illinois Governor's Pollution Prevention Award

additional benefits for Navistar. Environmental compliance reporting has become much easier and no longer requires the environmental staff to make their annual scavenger hunt to find chemical use and inventory data.

Chemical use tracking allows problems to be identified quickly. Hendershott explains one example:

> One washer had a problem with the flow control for the automatic makeup. Every morning when they would turn it on it would overflow. Nobody saw anything, they'd turn it on and it would seem fine. Now we can track usage by machine . . . so it was a case where we could see that one month's usage on this particular washer was very high. So I started questioning people, going out there and looking into it. Next month it continued to be high and we were able to get the manufacturer in and found the problem. The usage went down below what it was before the problem.

Rudy Bernath, Navistar's plant chemist, added, "It would have probably gone on for years. Even if we found it when looking at the year-end numbers, we might not have pinpointed that machine—the whole department's usage would have been high."

Having an on-site representative means that Castrol can identify and resolve chemical related problems before they get out of hand. This has produced benefits for Navistar *and* Castrol. Navistar benefits from Castrol's ability to make needed process or chemical product changes quickly and effectively before serious problems develop. When initial concerns began to surface about the health risks of coolant additive diethanolamine (DEA), Castrol was able to phase in a substitute without interrupting production. When a coolant or cleaner is causing a problem on the line, a call to the on-site manager usually resolves the problem quickly. Bob Hendershott relates another story:

> We had this concern with a maintenance cleaner, a Castrol product. It has a high pH—12 or 13. It's hazardous but compatible with the coolants. So we all talk about it. What pH range are you looking for? Something around 10, 10.5. I go back to our corporate people to find what would be compatible with the coolant, be a good cleaner, but have a lower pH. We identify a new product. I bring the MSDS and Rudy [the plant chemist] takes it to his meeting and we review it. If it looks fine, we bring a sample of it in.

Jerry Mittlestaedt commented on how well Castrol anticipates problems. Some of the additives they use contain Toxic Release Inventory (TRI)-reportable materials. Mittlestaedt was becoming concerned because "some of these chemicals were not over the reporting limit, but they're getting up there." When he brought the concern up with Hendershott, he found that "Bob's already got a plan in place and working on it. I couldn't believe it."

Castrol benefits from the ability to provide greater customer satisfaction. In a traditional relationship, the sales representative might hear about a problem only when someone got mad enough to call. As Bob Hendershott relates, "Being at Navistar on a regular basis allows me to pull samples and do testing on a regular schedule. I identify problems as they occur, so problems do not happen very long before I intervene. Therefore, problems don't continue for two weeks until someone calls and says, 'Castrol, your product isn't working. We got two weeks of bad parts because of you guys'." Jerry Mittlestaedt agrees. "With a quick Castrol response, we nip the problems in the bud." For a summary of benefits for Castrol, see Table 15-3.

Castrol has increased the size of the Navistar account not by selling more chemical but by providing additional chemical solutions to the customer's problems. In addition, Castrol also has immediate access to a real-world production environment to identify new chemical needs and applications as well as a beta-test site for new products.

An unexpected but valuable benefit for both Navistar and Castrol has been public recognition of their efforts. Based on the work of the Navistar-Castrol team, the Melrose Park plant was awarded the Illinois Governor's Pollution Prevention Award. The Melrose Park plant is recognized as a community and industrial leader in efforts to reduce waste.

Table 15-3.
Selected Benefits for Castrol

- Expanding chemical and service footprint
- Customer loyalty
- Opportunities to expand to other Navistar plants
- Inside information for market research and development
- Product R&D test opportunities
- Experience that could be used to obtain new accounts

THE PROBLEMS

"Castrol had to earn everyone's trust over time,"
—BRIAN NORDMAN, *first on-site Castrol representative*

No relationship is problem-free. Successful relationships come from the ability to work through problems and to learn from the experience.

Production problems are not uncommon. "I occasionally get calls at home," said Bob Hendershott, "from supervisors or process engineers on the floor. There may be parts going through a washer, then all of a sudden they are getting some residue on them, or rust. I am glad that they call because I can go out there and help assess the situation. If there is a day or so lag, the customer is making a lot of parts that may need to be reworked. If I can take care of it right away, I save a lot of trouble down the line."

Rust is an ongoing concern for Navistar, and Castrol's chemicals are designed to help prevent it. Liability for a rust problem falls to whomever is responsible. Rust prevention is also one of the performance specifications included in the formal contract. "If there are occurrences of rust that are within Castrol's control to prevent, then we are liable for it," explained Hendershott. "If it is something that is mechanical, outside of our control, then we are not liable for it. There are provisions for this in the contract."

But establishing responsibility is not always easy. As Jerry Mittlestaedt recalled, "There was one case of rust where the cause could not be identified. Castrol took responsibility and paid for it." Rudy Bernath added, "It wasn't really proven, whose fault it was, whether it was the product or inappropriate application of the rust preventative. It was never really proven, but Castrol, being a good sport about it, had the parts reworked."

The ten-year relationship has seen numerous problems, but all have been satisfactorily resolved. Though liability was a significant concern at the beginning of the relationship, it's rarely a problem today. In fact, the Melrose Park personnel have a hard time sympathizing with the liability concerns of many companies who perceive it as a barrier to implementing a Shared Savings relationship. "Why are they so afraid of liability?" asked Mittlestaedt. "Isn't that part of doing business?" Hendershott reflected, "You have to take responsibility. If some chemical is being applied or used improperly, document it in a letter or a memo. It's the same process I would use in a sales situation. If a company is not taking responsibility for its chemicals and an accident happens, it deserves to be liable for that." All agreed that in both traditional sales and Shared Savings relationships, both companies must take responsibility for their products and decisions.

THE FUTURE

Recently, the Navistar-Castrol contract switched from a fixed-fee format to a variable-fee format, where Navistar pays for chemicals by the pound or gallon. What appears to be a step backward actually represents a maturing of the relationship. Coolant usage had dropped so low that the monthly fee appeared to be too large relative to coolant volume. The contract was modified to pay Castrol by the pound for chemicals, though the chemicals continue to be managed on consignment at the Navistar plant—that is, Castrol owns them until they are used by Navistar. In addition, Castrol receives a management fee for chemical management services.

The change continues to bring progress in chemical use reduction. Because Navistar now pays for chemical by volume, it is even more sensitive to efforts to reduce waste. In addition, both parties supported a gainsharing clause in the contract. Ideas from Castrol personnel that can reduce costs or improve production for Navistar are rewarded through a sharing of the financial benefits. Thus, the Shared Savings spirit of the contract continues in the relationship.

Another recent change is the expansion of Castrol's chemical footprint to include rust preventatives and lube oils. Though services in these areas are more limited, the additions indicate the success of Castrol's chemical management efforts and the value they create for the customer.

With all the reduction in chemical usage, it might appear that the need for the Castrol representative in the plant has declined. Jerry Mittlestaedt responded immediately to this issue, "No, no. If he [Hendershott] left, it would all go to hell in a handbasket. This is a 1.5-million-square-foot facility. We make over two hundred engines a day. It's cost-effective to have Bob here every day."

"Somebody has to watch what's going on," added Rudy Bernath. "You've really got to watch. If you don't control the system, it will go all out of whack. Water won't get pumped out, the wrong product will get added, parts will get rusty." Jerry added the final word, "It saves us tons of money and improves the environment. This whole thing is just good business."

16

General Motors Truck
and Bus Plant

Shared Savings Case History
General Motors Corporation
Bus and Truck Assembly Plant
Janesville, Wisconsin
and
BetzDearborn

Fritz Benton thought it was just another review from corporate. Benton had been the BetzDearborn on-site chemical manager at GM's Janesville plant for almost five years, and he had been through many corporate reviews. That day his district manager, area manager, and even the corporate field advisor had arrived at the plant. But this was not to be just another review. When the team had assembled with an array of GM personnel, they had a surprise for Benton. He was presented with BetzDearborn's Customer Satisfaction Award. Benton had been nominated by his superiors, and many GM personnel had been involved in the interviews leading to the award. He considers it one of his proudest accomplishments.

What was most unusual about this event was that Benton earned the award by helping GM *reduce* the amount of chemicals needed from BetzDearborn!

SUMMARY

The General Motors (GM) Truck and Bus plant in Janesville, Wisconsin, has had a Shared Savings relationship with BetzDearborn since 1991. The contract covers a wide array of chemicals and systems including wastewater treatment, paint detackification, power house, maintenance paints, and solvents. BetzDearborn serves as a Tier 1 supplier for all these chemical services at the GM plant. The services are based on a variety of unit-pricing strategies. Known as the Chemical Management Program (CMP), it has produced substantial savings for GM–Janesville through reductions in chemical use and improved chemical management.

Both GM and BetzDearborn have experienced many benefits in other aspects of their operations as well. Improved chemical management in the plant has meant that GM personnel can focus on production while chemical-related headaches and costs decline. There are fewer chemical products in the plant, inventory is better managed, chemical use is tracked in detail, and health and safety conditions have improved. For BetzDearborn, revenues continue to grow even while GM is able to reduce chemical use. Perhaps, most importantly, BetzDearborn has won the opportunity to renegotiate its contract with GM rather than rebid—a sizable dividend of customer loyalty.

THE JANESVILLE ASSEMBLY PLANT

General Motors has built trucks, buses, and other vehicles at the Janesville assembly plant since 1922. Today the plant operates two assembly lines. One produces light-duty vehicles; such as the Chevrolet Suburban and the Tahoe; the other produces a variety of medium-duty trucks and buses. The production facility occupies over 3.5 million square feet and employs approximately five thousand workers.

BETZDEARBORN

For many years, Dearborn USA was one of the nation's leading companies specializing in water treatment chemistry. It served as a supplier of water treatment chemicals to many GM plants as well as those of other automakers. Dearborn responded to GM's drive to implement CMP programs with chemical suppliers and bid competitively for a number of Tier 1 contracts. At the time Dearborn won the Janesville contract, it was owned by W.R. Grace. Later, it was purchased by Betz, another nationwide water treatment chemical supplier. The sale did not change the GM-Dearborn relationship at Janesville.

Fritz Benton, the CMP on-site manager for BetzDearborn, has never been in chemical sales, and that's an advantage. Instead, he had years of experience in industrial water treatment. Benton was hired specifically for the Janesville contract. The hiring process involved both Dearborn and GM personnel. In fact, Benton was interviewed four times by Dearborn and twice by GM before being offered the job.

He is comfortable with a job that requires him to apply his expertise to help GM reduce chemical costs and improve chemical performance rather than to sell chemicals. BetzDearborn supports this strategy by paying Benton a salary plus performance bonuses rather than a commission on the volume of chemicals sold.

THE GM-BETZDEARBORN RELATIONSHIP

We were assuming responsibility for our supplier's product. If we could get our suppliers to assume that responsibility, it could save us a lot of money.
—MARK OPACHAK, *Worldwide Facilities Engineering, General Motors Corporation*

The Contract

The CMP contract at Janesville covers most nonproduction (indirect) chemicals, including water treatment chemicals, solvents, commodity chemicals, lubricants, and maintenance paints (see Table 16-1). However, the focus of the contract is not so much chemicals as the systems that use chemicals—paint booths, wastewater treatment, airhouses, and many more. BetzDearborn shares responsibility for the performance of these systems, including ordering and maintaining chemical inventories, managing chemical use and disposal, researching system improvement, facilitating EHS compliance, and training employees. One of the most important responsibilities of the on-site CMP manager is communication, particularly in helping to coordinate efforts among manufacturing, maintenance, wastewater treatment, and other departments.

Table 16-1.
The CMP Contract

---

1. **Chemical footprint**—water treatment chemicals (powerhouse, cooling towers, wastewater treatment, air houses), paint detackification and booth maintenance, lubricants, maintenance paints, commodity chemicals, purge solvents
2. **Financial relationship**
   - Fixed fee per unit of production (unit pricing)
   - Unit of production varies by type of chemical—e.g., production of vehicles, wastewater, steam, etc.
   - Management fees for selected services
3. **Risk/Reward**—mechanisms for quarterly adjustment of unit price due to significant production or operating changes, significant raw material price changes, other unforeseen factors, or incorrect information from GM. Formula for sharing unusually large financial losses.
4. **Responsibilities**—GM's UAW employees provide most of the hands-on work, including chemical changes and additions. BetzDearborn's activities include:
   - Purchasing and inventory control
   - Monitoring and coordination of chemical usage
   - Research and improvement of chemical performance
   - Chemical and regulatory testing lab
   - Ongoing reporting and communication
   - Product and process engineering development
   - EHS compliance and training
   - Continuous waste minimization
5. **Liabilities**—basic provisions for ownership of chemicals, prohibition of silicone-containing materials, and financial commitments. Liability associated with individual events is determined on a case-by-case basis.
6. **Performance requirements**
   - Unit prices to be steady or declining
   - Targets for overall plant savings, including an annual savings equal to 5% of the value of the contract
   - Savings may involve contributed R&D
   - Use of drums to be minimized
   - System performance limits (water discharges, chemical concentrations in boiler water, etc.)

The CMP contract employs a combination of techniques to create the right financial and operational incentives for both BetzDearborn and GM. First, it specifies a series of fixed fees (unit prices) for chemical-specific management services. For many chemicals, these fees are specified on a dollars-per-vehicle basis. However, because consumption of some chemicals is only remotely related to the rate of vehicle production, other performance measures are also used. For example, services related to power house chemicals are paid per million pounds of steam produced. Services related to demineralizing water are paid per thousand gallons of demineralized water produced.

Second, management service fees are paid to BetzDearborn for extra services such as chemical tracking and health and safety training programs. These are typically annual fees that do not vary with production or chemical use.

Third, several performance targets are specified. Unit prices for chemical services are expected to hold steady or decline over time. In addition, GM requires that cost savings of at least 5% of the value of the contract are achieved annually. However, this need not be only in the form of reductions in chemical costs. In fact, many of the most significant cost savings have come from improvements recommended by BetzDearborn but unrelated to the chemical systems managed by Fritz Benton.

BetzDearborn serves as the Tier 1 supplier, tapping Tier 2 suppliers for most solvents, commodity chemicals, lubricants, and maintenance paints. A similar cost-per-unit system is used by BetzDearborn to pay these Tier 2 suppliers, and several of the Tier 2 suppliers provide on-site chemical managers as well. Though these suppliers work for BetzDearborn, GM is involved in the process of selecting and approving Tier 2 suppliers and their products. On-site representatives from these suppliers often work directly with GM employees. However, all new product ideas, as well as all prospective suppliers, must be approved by BetzDearborn first. GM normally works only with ideas, products, and suppliers brought to them through BetzDearborn.

BetzDearborn is responsible for a full array of chemical management services from inventory management to waste minimization (many of these activities are summarized in Table 16-1). One of the many important activities is chemical purchasing. All chemical purchasing at the Janesville plant must be coordinated through BetzDearborn. GM employees who need a chemical cannot make independent purchases. The BetzDearborn on-site manager is expected to respond promptly to any employee's request, yet follow GM's chemical approval and acquisition procedures. This approach has eliminated the confusion over the number, type, and safety of chemicals used at the plant as well as the duplication of purchases and the buildup of overstocked and outdated inventory.

While basic chemical management activities, such as purchasing and inventory management, dominated the early years of the contract, chemical research and development (R&D) activities have more recently become an important component of the CMP relationship. From reducing paint detackification sludge to new paint booth coatings, BetzDearborn and the Tier 2 suppliers develop dozens of new product and process ideas each year. The supplier's on-site managers have become more valued as sources of chemical expertise than as sources of chemicals. Fritz Benton provides recent examples:

> GM Maintenance employees were having a problem with algae growth in the drinking fountains. They came to us for some of our biocides, but this was a problem they could solve with the proper use of bleach. It was a much simpler solution and saved them money as well. They also had a problem with some of the airhouses—a fishy smell. They asked us to help solve the problem. From water samples, the BetzDearborn labs were able to identify the organism and recommend an appropriate biocide. We also were able to recommend changes in procedures to minimize recurrence of the problem. Another problem was found in the fluid filling area of the line, where floors often got slippery. Working with Sherwin-Williams, our Tier 2 maintenance paint supplier, we were able to bring in a floor coating that retained its friction, yet was easy to clean and did not degrade from contact with the fluids.

The People

The BetzDearborn on-site manager and GM personnel enjoy an excellent working relationship. Fritz Benton, BetzDearborn's on-site CMP manager, reports directly to Mike Merrick, senior environmental engineer for the Janesville plant. However, Benton has regular contact with a wide variety of plant personnel, from senior management to hourly workers. Benton and other on-site supplier representatives serve on the Chemical Management Committee. In addition, Benton is expected to meet for regular updates with the booth cleaning supervisor, booth cleaning hourly workers, pipefitter supervisor, and the power house chief engineer. Benton produces minutes for many of these meetings. In addition to Benton and two part-time BetzDearborn technicians, many of the Tier 2 suppliers provide part-time on-site management personnel. Together, over 5,000 man-hours of chemical expertise are provided on site each year.

EVOLUTION OF THE RELATIONSHIP

Corporate Origins

The origins of the General Motors CMP program date from the mid-1980s, during the effort to win back market share from the growing tide of import autos. Measures to identify non-value-added activities led a group at GM to rethink the quality control procedures for incoming chemicals. Mark Opachak, one of the members of that original team, explains:

> We realized that each plant spent a tremendous amount of time and money checking and double-checking the quality of chemicals that were entering the plant. In some plants we spent well over $1 million a year. We were assuming responsibility for our supplier's product. If we could get our suppliers to assume that responsibility, it could save us a lot of money. We needed to make it in the best interest of the supplier to take that responsibility. We needed a partnership.

To buy into that partnership, GM realized suppliers would need a direct stake in any benefits it produced. The automaker offered this in the form of unit pricing—paying the supplier a fixed fee per unit of production regardless of the volume of chemical used. Combined with the supplier consolidation efforts underway at the time, this idea developed into a powerful program that changed the way GM handled chemical supply throughout the corporation. GM could focus on its core business, building cars and trucks, while the chemical suppliers could focus on their core business, chemical management.

CMP at Janesville

At the time, in Janesville, Wisconsin, Dearborn was a minor supplier of wastewater treatment polymers to the Janesville Assembly Plant under a standard chemical supply contract. However, when the corporate CMP reached the Janesville plant in 1990, Dearborn had already demonstrated its performance abilities and was able to submit a competitive bid to provide chemical management services.

In 1991, Dearborn was awarded the plant's two-year chemical management contract, covering water treatment chemicals for wastewater treatment, powerhouse, cooling towers, and welder water systems. It also included paint detackification for the medium-duty truck side of the plant. In 1992, the program was expanded to include paint detackification for the Suburban and Tahoe line as well.

Expansion

Dearborn won a three-year extension of the contract in 1993, expanding the footprint to include additional paint detackification systems and paint booth maintenance products. This was expanded again in 1994 to include lubricants, commodity chemicals, and maintenance paints as well as management of the paint line purge systems.

Most of the chemical systems added to the CMP contract were beyond the scope of water treatment and did not involve chemicals produced by BetzDearborn. As a result, BetzDearborn's Tier 1 responsibilities expanded substantially to include managing Tier 2 suppliers who oversee commodity chemicals, maintenance paints, lubrication chemicals, and purge solvents. Many of these suppliers, in turn, manage supply contracts with Tier 3 suppliers.

Because of BetzDearborn's demonstrated chemical management expertise, GM added a chemical management fee to the contract. This fee pays for overall chemical use tracking and reporting, certain health and safety programs, and other management activities. One recently added responsibility involves extensive upgrades to chemical safety labeling in the plant; another is collecting data needed to document volatile organic compound (VOC) reductions as well as for the Environmental Protection Agency (EPA) SARA Title 3 reporting requirements.

THE BENEFITS

This gives us a huge competitive edge.
—FRITZ BENTON, *on-site chemical manager, BetzDearborn*

Benefits to GM

Most companies struggle to keep annual increases in chemical purchase expenditures to a minimum, but under the CMP, there has actually been an 8% *decrease* in chemical costs per vehicle since 1993, the base year for the contract at Janesville (see Table 16-2). This cost factor includes the

Table 16-2.
Selected Benefits for GM–Janesville

- Substantial overall savings
- 8% decrease in chemical costs with significantly expanded services
- Improved inventory control; reduced inventory costs; product consolidation
- Training and other programs to improve health and safety protection
- Chemical tracking for easier compliance reporting
- Reduced VOC emissions and sludge disposal
- Reduced downtime and labor cost for sludge clean-out
- Many other improvements that reduce labor overtime, improve process efficiency, improve product quality, and reduce rework

management fees paid to BetzDearborn for their expanded services. The savings are the result of improvements made by BetzDearborn and the Tier 2 suppliers.

Though GM has requested that specific financial data not be disclosed, the program has produced substantial savings in chemical-related costs at the Janesville plant. Much of this is due to reduced chemical purchase costs, but more than half resulted from reducing hidden chemical costs. For example, the plant saves substantial amounts each year in inventory management costs and in equipment and computers provided by the suppliers. Because the contract requires GM to generate only one purchase order per year, it sees significant savings in personnel time related to the chemical acquisition process.

Other projects have also helped to reduce the chemical burden at Janesville. Paint detackification systems typically produce large amounts of sludge. Periodically, sludge settling tanks must be cleaned out. This requires that manufacturing operations be shut down and is often scheduled for holidays or other down times. BetzDearborn recommended changes that reduced sludge clean-out time by 80%, saving substantial labor costs. Also related to the paint sludge system, BetzDearborn did laboratory research on paint sludge chemistry, saving GM consulting expenses. An audit program by Tribol, the Tier 2 lubricants supplier, uncovered three malfunctioning lubricators, resulting in a substantial annual savings for GM. Many of the suppliers provide employee training in the proper use of chemicals as part of their chemical management services.

Yet more activities have reduced chemical-associated costs, even though the savings are difficult to quantify. For example, the number of chemical products has been reduced significantly in all areas, including a reduction of over 50% in the number of maintenance paints used in the plant. Product consolidation reduces the cost of product approval, health and safety training, and environmental reporting. Chemical inventories have been reduced, including a reduction of over 78% in maintenance paint inventory. Reduced chemical inventories decrease the risk of chemical spills and other chemical-related liabilities. Chemical use audits by BetzDearborn and the Tier 2 suppliers have improved chemical use efficiency while reducing emissions and employee exposure.

GM has also experienced benefits unrelated to the chemicals covered by the CMP contract. For example, the cotton covers used to shield robots from oil and dirt were thrown away after they were soiled. Working with BetzDearborn's contacts in the industrial laundry field, Fritz Benton set up a program to wash the covers instead of disposing of them. Now covers can be used as many as six times before wearing out, producing a significant savings for GM.

Benefits to BetzDearborn

The greatest benefit to BetzDearborn is the competitive advantage the GM contract conveys in the highly competitive chemical industry (see Table 16-3.). "This gives us a huge competitive edge," remarked Fritz Benton. "There aren't many companies who can offer this kind of service."

One tangible outcome of that edge is the opportunity to renegotiate the Janesville contract rather than rebid it. Explained Benton, "We've done a good enough job here that they're pretty happy with us." Rebidding a

Table 16-3.
Selected benefits for BetzDearborn

- Revenues from cost reductions
- Expanding chemical and service footprint
- Customer loyalty and the opportunity to renegotiate rather than rebid the contract
- Opportunities to expand to other GM plants
- Inside information for market research and development
- Product R&D test opportunities
- Experience, including valuable chemical tracking experience, that can be used to obtain new accounts

contract is costly for both sides. BetzDearborn successfully bid the contract in 1991 and again in 1993. Renegotiation means the company can put its time and effort into performance and product improvement rather than bid preparation.

Another tangible outcome is expanding business opportunities. Within the GM plant, BetzDearborn's footprint has expanded yearly, both in terms of chemical systems covered and services provided. The CMP also has provided the opportunity to expand into other GM plants and to use the experience at GM to win contracts with other companies.

BetzDearborn has been able to secure management fees on top of existing unit prices for certain value-added management services. Though these cover a wide array of services, chemical use and emissions tracking has become a prominent management-fee component of the program. Benton notes the competitive value to his company:

> Chemical usage, VOC tracking, SARA Title 3 reporting—these are services many companies need. Yet a lot of companies do not have the staff to track and report these as well as they would like. Their environmental engineers are swamped with other duties. Chemical use is our business, and we have aligned ourselves with some other companies to prepare a total tracking system if our customers need it.

BetzDearborn developed a new computer program to provide these data and information services for customers as well its as own chemical operations.

THE PROBLEMS

"You have to maintain a cordial working relationship," explained Benton. "That's how you work through the problems." A good example was how GM and BetzDearborn handled a chemical spill:

> One time, a GM employee accidentally put a forklift tong through a chemical tote. Of course, we responded to the spill and managed the cleanup. But we lost a lot of chemical—and in this plant, it's our chemical. There could have been a lot of arguing about fault and responsibility. But both GM and BetzDearborn know that their relationship is more important than the cost of the lost chemicals. The GM people noted a recent accounting error that had overcredited the BetzDearborn account. The two dollar amounts were about equal, so both sides called it a wash. It has to

be a cordial and fair relationship. There has to be a commitment to work through problems, not take advantage of them.

The commitment to work through problems and maintain long-term relationships with chemical suppliers is evident even at the corporate level in GM. Mark Opachak, with GM's World Wide Facilities Engineering Group, described this commitment:

> Look at the Chemical Management Program as a marriage. Our contract has a divorce clause, but in a healthy marriage you don't get married with plans for a divorce, and you don't stay married out of fear of divorce. Fear is not a good motivating factor. There has to be a positive, fair relationship with a desire to be together. The divorce clause is one millionth of the contract's importance. As far as we are concerned, a CMP contract is forever! Divorce is not an option.

Yet coping with change can put stress on any marriage. While change is inevitable and good for a company, it can pose challenges for a chemical management program. GM has made recent expansions and equipment changes at Janesville. Why is this a problem? Benton explained, "It's ideal when you have good baselines and everything stays the same. You can compare your improvements easily to that baseline. But when things start changing out there, how do you know you are doing better? We're working on this together, but to tell the truth, we don't have all the answers, yet."

Getting the program started at Janesville was not without its difficulties, even though it had strong support from corporate and plant management. Winning over the personnel who would be directly affected by the program was a priority. Bill Knick, chief engineer for the power house, recalls his own experience:

> We understood the old way of doing business—chemical sales. Going to just one supplier, that didn't sound right. And I thought I would lose control of decisions around here. But it hasn't turned out that way at all. I'm still in charge and I can monitor performance even better than before. The supplier brought in someone who really understood boilers and understood my problems. In fact, I was involved in picking the supplier. They have provided free consulting that has saved us money. I'm pleased with the new program.

Linda Little, environmental engineer at the Janesville plant, put it this way: "Management, unions, environment, health and safety staff—they all have to buy-in. That takes good on-site people from the supplier. They have to be *people people.*"

Mike Merrick, senior environmental engineer and CMP coordinator at the Janesville plant, noted that the program must be implemented carefully, involving everyone along the way. "GM has a nineteen-step process that is used to bring CMP in at any plant. From prequalifying suppliers to writing performance specifications, it is important to keep everyone involved."

Susan Kelsey, with GM's corporate World Wide Facilities Group, explained the corporate view. "To implement CMP successfully requires a team approach. People have to be involved. Failure to use a team approach is the number-one barrier to successfully implementing CMP at a plant."

Another problem can arise when the CMP Tier 1 supplier interfaces with other suppliers outside the CMP umbrella. For example, at Janesville, body paint is not under the CMP. Paints are supplied by another supplier on a traditional dollar-per-gallon basis. However, the actions of the paint supplier can affect the chemical management program. For example, tests of a new paint product require multiple runs through the spray guns, with flushing of the hoses and guns between runs. Both the purge solvent used in flushing the guns and the waste generated by the tests fall under the responsibility of BetzDearborn and the CMP contract. The added costs require documentation in order to seek compensation from GM. However, instead of responding to these ongoing conflicts through legal or contractual means, BetzDearborn responded "cordially." Its personnel moved the office of the purge solvent representative into the paint department, next to the office of the paint supplier, and worked to establish a closer relationship among all the parties. Problems related to the paint testing and cleaning declined significantly.

A potential problem that BetzDearborn and GM take seriously is job security. They do not consider this program to be a job-reduction strategy. They do believe their actions can relieve headaches from existing jobs and help redirect employee time to production and process improvement. Benton related the following story as one of his favorite examples of how this program can improve employees' work lives:

> We were going into another plant with a CMP contract that included the
> wastewater treatment plant. The chief operator was getting ready to retire
> at the time of the contract. However, after we came in and provided ad-

vice and helped make a lot process changes, he said his job had improved so much that he decided to stay another two years!

BetzDearborn also takes union relations seriously. "We were very careful not to take any union work away," noted Benton, though he admitted that sometimes workers move to other jobs when the chemical activities they were performing are no longer needed. He continued:

> We try to be an asset to the union members by keeping them informed. I talk with them about upcoming changes and work to make the changes smooth and manageable. Or we may be able to help them with other problems. They were using a cleaner that required extensive personal protective equipment. They came to us for suggestions and we were able to find a product that works just as well, and they don't have to suit up anymore.

THE FUTURE

The future of CMP at Janesville looks bright. BetzDearborn and the Tier 2 suppliers already have dozens of improvement ideas in the works. These include a plan to install a bulk purge solvent tank, which would save money on materials handling; the use of in-line purge solvent flow restrictors to further reduce solvent waste and air emissions; and placing the lubrication supplier, Tribol, in charge of the lube cycle program for 360 body shop robots.

A major addition to the contract that both sides are looking forward to is the implementation of gainsharing. Under this program, ideas from BetzDearborn in areas beyond chemical management that result in savings to GM translate into additional revenue for the suppliers as well as benefits for GM. This is intended to promote the implementation of creative ideas of the sort BetzDearborn has already generated. Through the gainsharing program, documented savings from such ideas will be shared by GM and BetzDearborn. This approach is consistent with GM's vision of the CMP concept. "Gainsharing is where the real benefit lies for GM," explained Mark Opachak. "It's that extra pair of eyes, that fresh look, that an experienced supplier brings to the plant that can help us solve the big problems."

Fritz Benton saw no end to the GM-BetzDearborn relationship. "This program just continues to grow," he said. "There are more opportunities everywhere we look. I love my job!"

17

Ford Chicago Assembly Plant

Shared Savings Case History
Ford Motor Company
Chicago Assembly Plant
Chicago, Illinois
and
PPG

"Don't unpack your bags, you won't be staying that long," was the comment that Tim Gillies heard, in one form or another, from people all over the plant. Gillies, a field engineer with PPG, had just arrived at the Ford Chicago Assembly Plant to implement a Shared Savings program. "Those were tough times," he recalled. PPG had replaced a popular supplier at the plant, and many people were suspicious of this new program. They did not trust the new concept—it just didn't make sense.

That was 1993; today, not only has PPG stayed but its relationship with the Ford plant and its personnel has flourished. The Shared Savings program that Tim Gillies implemented worked so well that today it covers the majority of chemicals used in the Ford Assembly Plant.

SUMMARY

Ford Motor Company has been using the Shared Savings concept at the Chicago Assembly Plant since the late 1980s. Its relationship with PPG as a major Shared Savings supplier dates back to 1993. PPG is responsible for most of the chemicals in the plant, with the exception of chemicals that become part of the product (paints, sealers, and lubricants). The two Shared Savings programs, known as Total Fluids Management and Total Solvents Management, have produced dramatic environmental and cost benefits, including a 57% reduction in volatile organic compound (VOC) emissions and a 27% reduction in wastewater treatment sludge, while maintaining or reducing chemical costs.

Both Ford and PPG have experienced many additional benefits; product finish quality, environmental performance, and employee health and safety have all improved as well. For PPG, the value of the Chicago Plant account has increased threefold over the short life of the program.

THE CHICAGO ASSEMBLY PLANT

In 1914, Ford Motor Company began assembling Model-T touring sedans in Chicago. After moving operations to the present location in 1924, the plant underwent many expansions and renovations, including a $178-million-dollar renovation in 1985 to prepare for Taurus and Sable assembly. Today the plant occupies 2.5 million square feet and employees over three thousand hourly and salaried workers. Production exceeds 250,000 vehicles per year.

PPG

PPG Industries began in the late 1800s as a plate glass manufacturer and expanded into the chemical industry as it grew. Today its predominant chemical market is industrial paints and coatings, but its purchase of Chemfil enabled the company to supply additional chemicals and chemical services. The Chemfil division of PPG now plays a significant role in marketing and servicing the Shared Savings market.

PPG perceived the potential value of a comprehensive Shared Savings program years ago. Though other companies were offering Shared Savings contracts on selected chemicals, PPG developed a program to provide such benefits on the full array of chemicals needed by major manufacturing facilities. Today, PPG is one of only two chemical suppliers authorized to provide Total Fluids Management at Ford facilities and currently manages programs at ten Ford plants. PPG understands that the Shared Savings approach to chemical management is an important business strategy.

THE FORD-PPG RELATIONSHIP

Ford Motor Company has said that it no longer
wants to be in the chemical business.
—TIM GILLIES, *PPG On-site Manager*

The Contract

The Shared Savings relationship with PPG at the Ford Chicago Assembly
Plant covers nearly every chemical in the plant (see Table 17-1). Paints,
sealers, and lubricants are the only major chemicals supplied outside the
relationship. The program between Ford and PPG is actually composed
of two separate contracts: Total Solvents Management covers all the sol-

Table 17-1.
The Contract

1. **Chemical footprint**—solvents, cleaners, surface treatments, water treatment,
   and most other plant chemicals other than paints, sealers, and lubricants
2. **Financial relationship**
   * Fee per vehicle for solvent management
   * Fee per vehicle for management of other chemicals
   * Fixed annual fee for selected chemicals unrelated to production
3. **Risk/Reward**—provisions for adjusting contract terms. This can cover large
   changes in chemical costs, large unexpected gains or losses, or other factors
   affecting contract terms.
4. **Responsibilities**—lists the responsibilities of Ford and PPG. Ford UAW em-
   ployees provide most of the hands-on work, including chemical changes and
   additions. PPG activities include:
   * Purchasing from Tier 2 suppliers
   * Inventory and distribution management
   * Container management
   * Quality assurance and maintenance oversight
   * Testing and lab analyses
   * Product and process engineering development
   * VOC emission reduction training
   * EHS studies and training
   * Process and waste problem solving
5. **Liabilities**—sets out an eight-step process to be followed in establishing the
   causes of a problem and the responsible parties, if any
6. **Performance requirements**—expected annual reductions in the fee per vehicle
   as well as a productivity reduction—a negotiated level of savings to be realized
   by Ford

vents used in the plant, while Total Fluids Management covers most other chemicals.

Both the Total Fluids and Total Solvents contracts are three-year agreements. Though the Chicago plant began the first Shared Savings program in the mid-1980s as a local program, today the basic elements of these contracts are negotiated at the corporate level of both PPG and Ford, with the contractual details being left to the personnel of individual plants.

PPG is compensated on a fee-per-vehicle or unit price basis. The fee is based on historical costs and volumes for both chemical usage and vehicle production. In turn, PPG, in cooperation with Ford personnel and production requirements, purchases all the contract chemicals for Ford, including many chemicals not manufactured by PPG.

Because PPG's fee is fixed per unit of production, the company works continually to improve Ford's chemical use efficiency. It can improve its own margins by reducing the amount of chemical supplied to Ford under the fixed-fee arrangement. Ford benefits by negotiating targeted annual reductions in PPG's per-vehicle fee. In addition, when PPG experiences a significant cost increase or reduction, the additional costs or savings are shared with Ford on a case-by-case basis.

Dan Uhle, environmental engineer at the Chicago assembly plant, described the dynamic incentives for the supplier:

> For the supplier, the greater the responsibility and the greater the contract value, the more they want their people to be in here managing their chemicals to be sure we aren't using more than we should to build a car. We're paying them the same price for every car that we build. So, when they start to control their usage to meet the same specifications—for example, cleanliness of the spray booths—if they can use less material, then they save money and their margin increases. It becomes more and more important for them to have management involvement in the plant.

In addition, PPG is expected to provide a productivity reduction equal to 5% of the value of the contract. This can take the form of any documentable savings to the Ford plant.

The majority of the chemicals provided under these contracts are not manufactured by PPG. Instead, PPG manages an array of Tier 2 suppliers that provide both chemicals and chemical services to the Ford plant. Many of these companies were suppliers to Ford prior to the development of the Shared Saving relationship and Ford wanted to retain them as suppliers when the Total Fluids Management program was implemented. Yet PPG, as a Tier 1 chemical manager, assumes full responsibility for the management and performance of all the chemicals under the contract footprint. It supports a full range of chemical services, from purchasing to waste reduction, for both its own chemicals and those of the Tier 2 suppliers (see Table 17-1).

The People

PPG maintains a full-time manager and a full-time service technician on site at the Chicago assembly plant. Many of the Tier 2 suppliers provide either full- or part-time on-site representatives as well.

Ted Camer, PPG's on-site chemical manager, works with many Ford employees daily, from the line production staff to the plant manager. He communicates routinely with the supervisors and union personnel responsible for the maintenance of production equipment. In fact, Camer considers his working relationship with these individuals to be one of the most critical aspects of his job. "Their attention to the functioning of equipment and the timeliness of their responses to our requests are key to the success of this operation," commented Camer. "As a result," noted plant environmental manager Dan Uhle, "cooperation of hourly workers with the PPG manager is even better than with Ford staff, even though he has no direct authority over their work. He takes the time to let them know how much he appreciates their efforts."

Camer also routinely participates on various teams in the plant. One team is headed by Greg Kohut, manager of manufacturing engineering. The team oversees the Ford-PPG relationship at the plant, resolving any problems or concerns that arise. Though the team scheduled frequent meetings when the relationship began, only quarterly meetings are now required. Instead, Camer's participation in the daily production process and production-oriented teams has become more important and financially rewarding. A recent example is Camer's involvement in a team that is working to make continuous improvements in surface finish quality.

The Ford/PPG relationship has evolved into a true partnership. The on-site manager knows that working in Ford's best interest is in PPG's best long-term interest as well. Camer explained, "If a problem comes up in the plant, even if it's unrelated to our chemicals, I try to help out. I have an interest in Ford and the quality of its products." In return, PPG has gained

Ford's trust and loyalty. It knows Ford is interested in more than just the contract price. As Ford's Dan Uhle put it, "A supplier who is trying to get a foot in the door may bid a bit less, but the supplier trying to get a foot in the door isn't offering the same total service."

EVOLUTION OF THE RELATIONSHIP

Detackification

The start of the Shared Savings program at Ford's Chicago Assembly Plant highlights the underlying power of these supply contracts. In the late 1980s, the plant was having problems detackifying paint overspray generated in the painting booths. The supplier of detackification chemicals blamed it on the painting operation and recommended additional chemicals at additional cost. This not only failed to resolve the problem, it actually made it worse.

Staff at the Chicago plant were aware of new programs being implemented at other Ford plants, where a supplier was paid a fixed fee per vehicle in exchange for meeting process performance requirements. This concept focuses a supplier on product performance rather than product sales. In addition, it provides a significant incentive for the supplier to reduce chemical usage and ultimately chemical costs. The Chicago plant decided to evaluate the new approach with the paint booth system at the Chicago plant. A supplier was selected to manage the *performance* of the paint detackification system.

The fixed-fee program worked very well. Detackification problems were significantly reduced and the plant experienced reductions in maintenance and operating costs as well as in waste generation. This convinced plant management that the approach could be used to improve performance and reduce costs in other areas too.

Total Fluids Management

At the corporate level, Ford was developing the Total Fluids Management concept in cooperation with PPG and other chemical suppliers. The program specified that a Tier 1 chemical supplier would receive a fixed fee per vehicle to manage a large chemical footprint (excluding paints, sealers, lubricants, commodity chemicals, and solvents). Ford would no longer purchase these chemicals, only chemical services, from the Tier 1 supplier. PPG received contracts to provide Tier 1 chemical management services for ten Ford plants.

PPG initiated the Chicago Assembly Plant contract in early 1993, beginning with management of the phosphate/bonderite system only. The first on-site chemical manager was Tim Gillies, who transferred from a successful Shared Savings program at the Chrysler plant in Belvidere, Illinois. His goals were to reduce supply, inventory, and management problems while working to improve chemical performance. Another priority was to earn the trust of plant employees. PPG replaced a competitor who had established a good working relationship with the production personnel at the plant. "It was difficult, but they [PPG] finally did it," remarked Uhle. "And since then, it's just been improving continuously. They got their foot in the door and were given a chance to prove themselves. Now that Tier 1 relationship has grown significantly."

Total Solvents Management

The Total Solvent Management program was started at the Chicago Assembly plant in the mid-1990s to significantly decrease VOC emissions. It covers all solvents used in the plant. Ford's positive experience with PPG's Total Fluids program was influential in helping the same company win the contract and expand its chemical footprint in the plant.

The solvent management program at the Chicago assembly plant is critical to Ford's operation in Chicago because the plant is located in an ozone nonattainment area. The plant was facing severe regulatory restrictions for VOC emissions. The prime performance goal of the Total Solvent program was VOC reduction. Putting solvent management on a fixed-fee basis gave the supplier a significant incentive to reduce solvent usage and associated emissions of VOCs. The program has been successful, reducing emissions by over 57% in the first eighteen months of operation, thus helping Ford avoid significant emission control investments.

Commodity Chemicals

Another indicator of the Shared Savings program's success at Ford's Chicago Assembly Plant was the modification of the Total Fluids Management contract to include commodity chemicals used in wastewater treatment.

Specialty chemicals used in wastewater treatment were already covered under the contract. Including all wastewater treatment chemicals allowed PPG to optimize treatment chemistry, reducing the generation of hazardous waste from the water treatment plant by over 25%.

Other Commodity Management Programs

Ford found that the same strategies used in Total Fluids Management could be applied successfully to other areas of the plant with other materials as well. Over time, the Chicago Assembly Plant has initiated a Total Waste program, a Total Filter program, and a Total Paint program. Suppliers receive a fixed fee per vehicle for their services and use contract terms similar to those for Total Fluids and Total Solvents. The financial incentive to reduce waste, along with the associated integration of the supplier's technology and services into plant operations, has made these programs financially successful for both Ford and its suppliers.

THE BENEFITS

The Total Fluids and Total Solvent Management programs have led to stable or decreasing costs for Ford in an area that had seen significant annual increases in previous years (see Table 17-2). The result is significant benefits in the highly competitive auto industry. Ford has experienced similar financial benefits from programs covering wastes, filters, and paints. According to Dan Uhle, "Certainly one of biggest benefits with this program is that we have been able to get the management help that is necessary for continuous training, continual improvement in waste reduction, and even

Table 17-2.
Selected benefits for Ford Chicago Assembly Plant

- VOC emissions reduction of 57% in eighteen months
- Reduction of wastewater sludge generation by 27%, saving over $50,000 per year
- Steady or declining chemical costs
- Improved product finish quality; reduced rework
- Improved inventory control; reduced inventory costs
- Improved health and safety protection
- Easier compliance reporting

in improving quality. We've been able to do this without raising costs, and in some instances the costs have gone down."

However, the benefits from the Shared Savings strategy extend well beyond the savings achieved on reduction of chemical costs. Perhaps the most significant benefit has been the reduction in solvent use with respect to VOC emissions. Ford knew that a good chemical supplier whose core business was chemical management could produce better results in a shorter time frame than it could working alone. "We felt we were making every attempt possible to reduce our level of solvents," explained Uhle, "but we're approaching it even more aggressively with the Total Solvent contract. PPG is helping us to reduce VOCs with much more vigor than we could have on our own." The performance of the Total Solvent Program clearly bears this out. With a 57% reduction in VOC emissions in the first eighteen months of the program, PPG has demonstrated the power of a Shared Savings arrangement.

Another important waste reduction benefit produced from the PPG relationship is the reduction in sludge from wastewater treatment. The switch to aluminum for selected body panels had resulted in a reclassification of this sludge as hazardous waste. PPG and Ford began an extensive research program to find ways to reduce sludge generation. In association with Nalco Chemical Company, the Tier 2 supplier of water treatment chemicals, they changed the chemical processes used in wastewater treatment, reducing not only sludge but also the amount of commodity chemicals used in the treatment process. Two related activities contributed to sludge reduction. A magnetic system was installed in the phosphating bath to collect and remove weld flash slag. This indeed reduced sludge, plus it improved the quality of the finish by removing these impurities from the system. A deluge car wash system was instituted to better remove impurities from the surface of the car before the phosphating step. In total, these activities reduced sludge generation 27% and have saved Ford over $50,000 per year.

Inventory costs have also declined. Ford avoids the inventory carrying costs (as the chemicals belong to PPG) and benefits from improved inventory management. Out-of-date and off-spec product wastes have declined dramatically, while the timely availability of chemicals has improved as PPG moves closer to a just-in-time (JIT) inventory system.

In conjunction with improved chemical inventory, chemical tracking has also improved, allowing variances in chemical use to be identified early and tracked to each process. This level of control has improved trouble-shooting while greatly simplifying environmental reporting. Dan Uhle explained the benefits from his perspective:

> Chemical tracking has improved environmental reporting substantially. That used to be the hardest part in doing our Form Rs—coming up with good chemical usage data. . . . Take solvents, for example. I had to look at what we bought and get records from the suppliers of all of the different materials that had VOCs in them. I had to make the assumption that the inventory at the beginning of the year was the same as at the end of the year. I didn't know exactly how much was scrapped and actually went out as waste paint solvent, where we had some recovery, or how much was actually emitted. So I had to make assumptions about all of that to the best of my ability, using engineering judgment. Now PPG keeps daily records of what is used and the VOCs emitted, and they not only do it by product but they do it by process. This program provides much more help in maintaining control than I've ever had.

With PPG providing a centralized chemical purchasing service, it has been easier to maintain health and safety control data as well. Again, Uhle explained the difference that the PPG relationship has made in the health and safety program.

> Ten years ago, we would get some materials in here that didn't have internal toxicology clearance. But it's different now. Now it's much easier to keep control over all the products. With PPG, if you use it, you've got to have a toxicology approval number. Once we have that, I have all the information on a database and I have 100% disclosure of what's in it.

The benefits for PPG have been numerous (see Table 17-3). The combination of customer loyalty and an expanding chemical management

Table 17-3.
Selected benefits for PPG

- Expanding chemical and service footprint, increasing the value of the account three fold
- Customer loyalty
- Inside information for market research and development
- Product R&D test opportunities
- Experience that can be used to obtain new accounts

responsibility inside Ford plants provides long-term revenue security. Building a close relationship with a customer is a key competitive strategy for PPG. "For PPG, the greatest benefit has been the special relationship we've been able to foster with Ford," said Ted Camer. "There is a continuing trust between the two companies that wouldn't be there if we hadn't gotten into this relationship. It's been very gratifying on both sides." Since PPG entered the Shared Savings program at the Chicago plant, the value of their account has increased threefold.

The link between profit and chemical services expertise also gives PPG a competitive advantage in the chemical supply industry. Tim Gillies, the first PPG on-site manager, pointed out that "this is a good deal for PPG. It's designed to make money while saving Ford money as well. Our fee is based on historical chemical costs. If I can cut those costs substantially, there's my profit."

What underlies the benefits for both companies is the ability of each to focus on its own core business. As Gillies put it, "Ford Motor Company has said that they no longer want to be in the chemical business." Or, in the words of Dan Uhle:

> This allows us to concentrate on what our specialty is—building quality automobiles and being able to sell those to the public—we certainly want to build what they want. We are managing the manufacturing business and passing responsibilities on to people who are more experienced in managing certain subsystems.
>
> I'll give you an example from my own experience. If we bring in a material to do a bench trial out at the water treatment plant, we call up the Total Fluids, coordinator and he takes on the responsibility. Before Total Fluids, I had to do everything. I had to go get a sample of the material myself, I had to go take it out to the waste treatment plant myself, and I had to make whatever adjustments were necessary to reflect the concentrations that we would see and document it all—and that was very time-consuming. Now the Total Fluids coordinator can do it much more efficiently and bring back the results. I have time to more closely review those results to determine what effect they're going to have on our business.

THE PROBLEMS

Implementation of the Shared Savings programs at the Chicago Assembly Plant has experienced surprisingly few problems. In part, this is due to its origins in a well-defined need at the plant: to bring paint detackification under control. Though there was initial resistance from some plant personnel, strong support provided from the corporate level contributed to the successful implementation of the Total Fluids and Total Solvent programs. Dan Uhle believes the many obvious benefits were also responsible for overcoming resistance:

> I think people could see the advantages of this relationship. The maintenance manager knew that without the on-site manager, a supplier would not share responsibility for the maintenance program. Certainly, Materials Handling could see the benefits of reduced inventory carrying costs and having less chemical in the plant to manage. Plus, they no longer had to be responsible for making sure the chemicals were here on time. In terms of production, having someone make sure the chemistry in the equipment is working lets us focus more on quality—and everyone's concerned about quality. Another set of eyes and a partner in the process has made significant progress in the quality of our products.

The initial concern of many employees about the Total Fluids program faded after the first few months of the contract. Though PPG replaced a popular supplier, it proved its value as well as the value of the new program. It was not long before many employees were asking that the program be extended to other areas in the plant.

There was some initial concern among hourly workers. "Some workers felt that they would be subject to two areas of management where they only had one before," explained Uhle. "They looked at the supplier's involvement as another layer of management that was trying to get them to do more." Experience quickly resolved those concerns. "I think everyone's attitude toward quality improvement and waste reduction has changed. Job security is one of their priority concerns, and they understand that if we can't sell a high-quality product to the customers at a reduced price, some other company will do it. The supplier can help in this."

Another potential problem that never materialized was legal liability. Though the contract contains provisions for handling liability, this has never been an issue in day-to-day operations. As Ted Camer put it, "I guess we understand the liability instinctively, so it's not something we spend much time thinking about."

Resolving problems that arise in daily activities is always a challenge in Shared Savings relationships and a test of the on-site personnel responsible for the contract. Ford and PPG use an eight-step problem-solving approach to find the causes of problems and resolve their differences. The process is critical to identifying the facts about the origin of the problem and resolving it, thus avoiding personal arguments that could damage the relationship.

It is also important to maintain flexibility in the inevitable but unexpected gains and losses in the process. Dan Uhle explained:

> In order to keep a good relationship, there has to be a balance of the positive and negative. It can't be a one-sided relationship. There have been times that PPG lost out. For example, we used more product than expected but we didn't know exactly where we used it. So Ford, in return, makes sure PPG sees some gains as well. As an example of that, we have made significant progress in block-painting our vehicles. Under the Total Solvent program, we are paying per unit for the purge solvent. Before, we purged every unit or unit and a half, on the average, because of color change. Now we're purging every three or four units, or even more than that, so the amount of solvents we are using has declined significantly. Now Ford didn't change their cost per unit, even though all of the facility-change costs were borne by Ford. That's an example of the type of positive thing that happens from PPG's involvement in the process.

THE FUTURE

The Shared Savings relationships at Ford continue to evolve. The Total Paint Management program is growing, as Ford's three key paint suppliers—PPG, DuPont, and BASF—enter into fee-per-vehicle contracts at individual plants. Paint is one of the highest-cost, highest-volume, and most critical chemicals supplied to assembly plants. The role of the paint supplier in overall chemical management is expected to grow, possibly moving the paint supplier into a Tier 1 position over all chemicals in the plant. The current Tier 1 suppliers in programs such as Total Fluids and Total Solvents would then become Tier 2 suppliers under the Total Paint contract.

Because DuPont is the paint supplier at the Chicago Assembly Plant, this could mean a change in the contract between Ford and PPG there. Whether DuPont becomes the Tier 1 supplier or not, PPG does not foresee great changes in its role. According to Ted Camer, "To us, our relationship with Ford here at this plant is such that it really doesn't matter either way. It's not going to change the excellent relationship that we have built."

CHAPTER

18

Chrysler Belvidere
Assembly Plant

Shared Savings Case History
Chrysler Corporation
Belvidere Assembly Plant
Belvidere, Illinois
and
PPG Industries

No one had done it before! Bob Conrad of Chrysler's corporate Paint and Energy Management group was very proud. Scratches or blemishes on the surface of a car after assembly are a major concern for Chrysler, as they are for any automaker. The original paint is cured by high-temperature baking, a process that cannot be repeated once the car is assembled. Traditionally, repairs are made using chemicals that emit large amounts of volatile organic compounds (VOCs).

Now all that had changed. Through extensive research, a method had been developed to use the original paints in a high-temperature baking process without risking heat damage to other components in the car. This not only produced a higher-quality product for the customer but it also significantly reduced VOC emissions. In addition, it reduced the volume of chemicals needed by the plant and saved Chrysler money.

Larry Petty was just as proud—and Petty works for Chrysler's chemical supplier, PPG! In fact, Petty had initiated the project and worked jointly with Chrysler to perfect the new technology, even though it meant fewer chemicals supplied to the plant. Actually, *because* it meant fewer chemicals supplied to the plant!

SUMMARY

The Chrysler Assembly Plant in Belvidere, Illinois, has had a Shared Savings relationship with PPG since 1989. Originally implemented to service phosphating and paint detackification systems, the program, known as Pay-As-Painted, now covers all chemicals and systems related to cleaning, treating, and coating autobodies. The process begins with cleaning the incoming sheet metal and ends with the final clear-coat finish applied to the car. PPG is paid a fixed fee per vehicle. To fully integrate quality into PPG's responsibilities, the fee is not paid until the vehicle passes a quality inspection leaving the final painting operation.

The benefits for both companies have been tremendous. Chrysler saved over $1 million in the first year alone. VOC and solvent usage were dramatically reduced. (The Belvidere plant has won *four* Illinois Governor's Pollution Prevention Awards.) PPG's chemical footprint in the plant has expanded significantly; Chrysler is a loyal customer.

THE BELVIDERE ASSEMBLY PLANT

Built in the early 1960s, the Chrysler assembly plant in Belvidere, Illinois, has produced a variety of vehicles for the Chrysler Corporation. Today the plant assembles over one thousand Neons each day.

PPG

PPG Industries began in the late 1800s as a plate glass manufacturer but expanded into chemical manufacturing, specializing in industrial paints and coatings. Today, PPG is the world's largest manufacturer of automotive and industrial coatings. Its coatings and resins division manufactures automotive primers and finishes, refinishes, adhesives, and sealants.

THE CHRYSLER-PPG RELATIONSHIP

It requires a whole different kind of thinking.
—BOB CONRAD, *Environmental Specialist, Chrysler Corporation*

Table 18-1.
The Pay-as-Painted Contract

---

1. **Chemical footprint**—all autobody surface preparation, treatment, and coating chemicals (excluding solvents)
2. **Financial relationship**
   - Fixed fee per vehicle painted to quality specifications
   - PPG owns chemicals until used
3. **Risk/Reward**—cost savings are shared between both parties during model year. Usage targets and fees may be adjusted downward at end of model year. Reconciliation procedure available to adjust for short-term excess usages.
4. **Responsibilities**—lists the responsibilities of Chrysler and PPG. PPG activities include:
   - Acquisition
   - Enhanced inventory and distribution management
   - Container management
   - Quality assurance and maintenance oversight
   - Testing and lab analyses
   - Product and process engineering development
   - EHS studies and training
   - Process and waste problem solving
5. **Performance requirements**—quality and performance specifications drive the agreement. PPG's fee is based on producing a specified paint finish.

The Contract[*]

Chrysler's relationship with PPG as a paint supplier goes back many years, but the innovative Shared Savings program the two companies developed is barely ten years old. Known as the Pay-As-Painted program, it is named after its most distinctive feature: PPG is paid per vehicle painted (see Table 18-1). The program is implemented by a cross-functional Pay-as-Painted team, which includes Chrysler staff from a variety of departments and PPG on-site personnel.

PPG is responsible for all the chemicals used to clean, treat, and coat the autobody. The process begins with cleaning the incoming sheet metal and ends with the final clear coat applied to the car. It also includes key quality steps such as phosphating and electrocoating prior to painting. These processes are critical to Chrysler's corrosion protection guarantee.

---

[*]Actually, Chrysler uses a one-year purchase order (PO) instead of a contract. The PO contains pricing information, but other terms of the agreement are contained in a lengthy set of standard attachments.

Bob Conrad, environmental specialist with Chrysler's corporate Paint and Energy Management group, explained the program:

> The way it works is that we pay our supplier, PPG, for chemicals and services every time we produce a good car out of the paint department. We don't own the chemicals; PPG is responsible for them. If there is a problem with a chemical, they take it back or do what's necessary to correct the problem. We don't own the chemicals until they have worked successfully on our product. The point is that we pay only for a quality finished product. We pay our supplier for products that are actually salable.

Larry Petty, on-site program manager for PPG put it this way:

> Instead of a vendor wanting to sell more product to a customer, now the vendor is working with the customer to optimize and reduce the excess in the system. The first year, at this plant alone, there was approximately a million-dollar savings in usages—because we're focused now on reducing costs. Some people might say that's a detriment to the vendor because he's losing business, but it's given us an opportunity to expand into other areas that we weren't in before.

Not only does the contract cover the process of coating the autobody, it also covers the full range of responsibilities required to manage each chemical from ordering and inventory management through waste treatment and disposal. A key component of this service is chemical data management. PPG closely monitors inventories and usages; the company developed and patented a software package to systematically analyze and compare projected and actual usages as well as costs. This computer program allows the Pay-as-Painted team to quickly identify paint usage variances.

Establishing the fee per vehicle is simple in concept. The fee is based on historical costs for each chemical divided by the number of vehicles produced. However, establishing this payment ratio for paint is complicated by variations in the cost of different paint colors. PPG's data management software collected the initial data needed to establish the baselines. The task is even more daunting in some of Chrysler's other plants, where hundreds of paint combinations may be applied to the vehicles.

The Pay-as-Painted contract is actually an annual purchase order (PO) with a set of attachments specifying standard Chrysler PO conditions and contingencies. Price (the fee per vehicle) can be readjusted annually to reflect underlying changes in chemical usage. In addition, a reconciliation mechanism is available to adjust for unanticipated increases or decreases in chemical use. Excess usage by Chrysler is paid for by Chrysler through this mechanism. It provides a financial incentive to minimize waste, even though the chemicals belong to PPG.

The fee per vehicle creates a strong incentive for PPG to help the Pay-as-Painted team reduce chemical use. Again, Bob Conrad explained:

> Our supplier makes money immediately if the Pay-as-Painted team can eliminate waste. As efficiency improves, we split those savings with them. Then, at the start of the new model year, the targets are reset. If we are using less of a material because of an effort made by either ourselves or our suppliers, the payment per vehicle is reduced.

A parallel program is known as the Solvent Management program, which covers all processes that could significantly effect VOC emissions. Though most of PPG's responsibilities are the same as in the Pay-as-Painted program, Solvent Management is not a Shared Savings relationship; instead, PPG is paid a management fee for its services. However, Chrysler and PPG plan to move the Solvent Management program to a cost-per-vehicle Shared Savings program in the coming years.

Larry Petty, PPG on-site manager, explained how the Solvent Management program works:

> The program places a strong focus on the volume of usages, location of usages, and monitoring each step in the process of how a solvent is used. The information is put together by PPG, producing a detailed guide for every product that is used. Then we start from the top down, with the highest VOC emitter. When Chrysler began the Solvent Management program with PPG, they wanted us to incorporate new products into the system that were safer for the environment, as well as to find ways to manage the day-to-day operations and practices in the plant without using any of the contributors to VOC emissions.
>
> We're constantly looking for ways to improve processes. For example, we helped Belvidere win the Illinois Governor's Pollution Prevention award back in 1992, even before the formal Solvent Management program. It was due, in part, to a paint stripper product that PPG introduced into the Pay-as-Painted team. We have made so much progress under Solvent Management since then that now that particular product is the largest VOC emitter that we have. What got us the award then is now the biggest chemical we need to replace—we've come that far.

Many of the chemicals covered in the two programs are not supplied by PPG. Instead, PPG serves as a Tier 1 supplier, contracting with Tier 2 suppliers and overseeing the management of these chemicals in the plant.

The People

One of the most unusual aspects of Pay-as-Painted is the team approach used at both the plant and corporate levels. "The team approach is what really makes this a successful program," explained Ernie Schmatz, Pay-as-Painted coordinator with Chrysler's Paint and Energy Management group. At the plant, representatives from the paint shop, maintenance, environmental control, finance, production control, and other departments, meet weekly with on-site PPG personnel. The group reviews usage and cost data, identifies problems, and works out solutions. A parallel corporate team oversees the Pay-as-Painted process across plants.

PPG currently has eight full-time people on site. Six work with the Pay-as-Painted program, each taking responsibility for different systems in the production process. Two people, out of PPG's Chemfil Division, work in the Solvent Management program.

Larry Petty is PPG's on-site manager for the account. He and his staff work extensively with Bob Conrad every day. However, they also work closely with the Paint Center manager, area managers, and the spray booth hourly workers. In addition, each PPG employee has responsibility for different autobody treatment and coating operations and must work closely with Chrysler staff involved in those operations.

EVOLUTION OF THE RELATIONSHIP

The point is that you pay only for a quality finished product. You pay your supplier for products that are actually sellable.
—BOB CONRAD, *Environmental Specialist, Chrysler Corporation*

Getting Started

In the late 1980s, PPG had a traditional supply relationship with the Belvidere plant. At the corporate level, Chrysler and PPG were discussing innovative ways of restructuring their relationship to involve PPG more closely in product quality and to reward the supplier for improvements in process efficiency. The program was designed to address Chrysler's desire to improve budgeting, avoiding the spikes in chemical expenditures that made planning and budgeting difficult. In 1989, the decision was made to begin with the phosphating and paint detackification systems at the Belvidere plant. The plant manager championed the program and provided the support it took to put it in place and make it work.

Tim Gillies was the PPG phosphating representative to the plant at the time. He explained the original program:

> The decision was made at PPG that our objective was to integrate ourselves better with our customers. We looked at this program as a way to become a true partner—but to do it, we knew we had to make some investment in our customer's plants. It was a commitment—we're in it for the long haul.
>
> We launched the program with phosphate and detack. We designed a cost per vehicle using historical chemical use and production data. The idea behind the program was for it to be advantageous for both parties. If last year it cost a dollar per vehicle to run the process and we could get it down to ninety-five cents, that was profit we could share.

The program worked. In fact, it was so successful in improving quality as well as generating savings that it was quickly expanded to include paint booth cleanup. Chrysler was working to reduce VOC emissions from paint booth cleanup and hoped the program would accelerate that process. The resulting VOC reductions were so dramatic they contributed to winning the plant's first Illinois Governor's Pollution Prevention Award. The plant has since gone on to win the award three more times.

Pay-as-Painted

In 1991, PPG and Chrysler began laying the groundwork for integrating the painting systems into PPG's contract and implementing the Pay-As-Painted reimbursement program that was originally envisioned. As Tim Gillies explained, "Incorporating the e-coat process was relatively easy. We understood the system very well, and every car gets treated the same. But the paint shop is different. There are so many different painting schemes and so many factors affecting the system. We needed extensive baseline data on all our materials."

Larry Petty designed a software package that allowed PPG and Chrysler to develop unit costs for their various vehicle painting options. In 1992, all painting systems were covered under the PPG contract and paid for on a per-vehicle basis.

The contract incentives were also fine-tuned to ensure that everyone had the right incentives to reduce waste and improve efficiency. Petty explained:

> In order for the program to succeed, you have to have incentives for the customer as well as the supplier. This program has built-in incentives that keep everyone focused. My accountabilities are to help maintain the program and to optimize it. If I let the program go excess, not only Chrysler is going to be dissatisfied but my people are going to be dissatisfied.

Again, the expanded program proved extremely successful, producing over $1 million in savings the first year of implementation. The plant manager who implemented the program has now moved on to another plant, taking the program concept with him. Several other Chrysler plants have implemented Pay-as-Painted as well. Yet implementation in other Chrysler plants has been gradual. In part, this is because of the time it takes to collect the accurate production data and prepare the plant personnel for the new relationship. But implementing such a radically different chemical supply relationship actually takes more than this. As Chrysler's Conrad put it, "It requires a whole different kind of thinking."

Solvent Management

The Solvent Management program evolved naturally out of Pay-as-Painted. The need for additional reductions in VOC emissions meant that Chrysler had to investigate innovative ways to maintain product quality and performance while minimizing VOCs. There is a strong link between solvent use and surface treatment operations associated with painting. For example, the use of solvents to flush paint hoses and guns is related to both the quality of the paint finish and the amount of paint waste generated.

However, the Solvent Management program does not employ unit pricing, the fixed fee per vehicle that is the revolutionary idea behind the Pay-as-Painted program. PPG's Petty explained why:

> When we went into the Pay-as-Painted program, we used three years of history to establish pricing. For the products in Solvent Management, there wasn't enough good historical data. Now we're establishing that data. In

addition, you've got to know the system and get it under control. If you establish pricing and you don't have the system in control yet, it's going to be impossible.

As a result, Chrysler and PPG are currently using a management fee program. Most of the solvents are provided through a Tier 2 supplier, not by PPG, and continue to be purchased by Chrysler on a traditional cost-per-gallon basis. PPG receives a fee to manage the chemicals inside the plant. In the process, PPG and Chrysler are studying the use of solvents, working to bring solvent systems under better control, and collecting usage and cost data. Petty noted, "The next step for the Solvent Management program is unit pricing."

THE BENEFITS

The first year at this plant alone there was approximately a million-dollar savings in usages—because we're focused now on reducing costs.
—LARRY PETTY, *PPG on-site manager, Chrysler Belvidere*

The Pay-as-Painted program produced over $1 million in savings in the first year of operation (see Table 18-2). Much of this was split between PPG and Chrysler. Once usage targets and fees are readjusted to reflect efficiency improvements, those savings accrue to Chrysler—forever.

Because PPG is not paid until a car passes a paint inspection, it has a vested interest in the quality of the painting operation, so while costs have declined, quality has improved. Chrysler already has a reputation as a low-cost producer in the industry, yet Belvidere has both the lowest-cost and highest-quality painting process of all Chrysler plants. If finish-related

Table 18-2.
Selected benefits for Chrysler's Belvidere Assembly Plant

- Over $1 million in savings from the first year of Pay-As-Painted
- Dramatic reductions in VOC emissions and other wastes
- Improved product quality; reduced rework
- Improved inventory control; reduced inventory costs
- Improved health and safety protection
- Easier compliance reporting
- Four Illinois Governor's Pollution Prevention Awards

quality problems should be found after a Neon is shipped, a team of Chrysler and PPG personnel perform the field investigation.

Two of the innovations that PPG helped initiate at Belvidere under the Pay-as-Painted program are the water-based paints and powder antichip paint. The water-based paints replaced solvent-based paints, reducing VOC emissions while maintaining finish quality. Powder-based antichip is applied to the front and lower body sections of the Neons to reduce chips and nicks. The powder produces a thicker, more chip-resistant coating while dramatically reducing VOC emissions. VOC reductions were achieved not only because powder antichip replaced solvent-based paint but because powder antichip does not require the use of solvent in ancillary maintenance operations, such as hose purging and booth cleaning.

Another innovative development was technique to improve the application of phosphate/bonderiteing. This system deposits a critical corrosion-resistant coating on the autobody using electrodeposition. A common problem associated with phosphate/bonderite is depositing sufficient material on high-corrosion areas while limiting the amount of material on other body areas with minimal exposure to corroding elements. PPG's innovations significantly reduced overdeposition in noncorroding areas. This saved money by requiring less material to provide the same level of corrosion protection.

In addition, the software developed by PPG to monitor chemical usage and cost—key variables affecting PPG's profitability—allows the Pay-as-Painted team to control the painting system at levels not achieved before the contract. According to PPG's Petty, "This system makes it very easy to monitor costs and usages. We track most systems weekly. We track paint daily. When we need to refine a system, we can even compare usages across shifts, or actually split a shift. When we study a process in that detail, that's where we can get the biggest reductions."

Most recently, the Pay-as-Painted team initiated a research program that has led to a proprietary technology to repair paint scratches or blemishes created during assembly, using the same paint and high-temperature baking methods used to apply paint to a car in the normal painting process. This innovation improves the quality of the paint finish and significantly reduces the VOC emissions associated with older repair methods. Finally, it saves money both PPG and Chrysler money. With these improvements, they've applied for *another* Illinois Governor's Pollution Prevention Award!

An important additional benefit of the program, but one that defies quantification, is that as the strength of the Chrysler-PPG relationship continues to grow, PPG is better able identify and meet Chrysler's needs, even to the point of customizing chemicals for specific plant applications. Bob Conrad summarized it this way:

> They know better than we do what the chemical capabilities are. We know better than anyone what our needs are. When we get together, we combine that knowledge. It's a lot better than someone coming in off the street and saying, "I can solve your problem." They don't even understand our problems! With this program, we've got someone who's not just coming in the door to sell us things. They are here to work with us to solve the problems that we have.

A powerful, underlying benefit of Pay-as-Painted is that both Chrysler and PPG personnel end up performing higher value-added activities. As Bob Godare, environmental manager at the Belvidere plant, pointed out, "We've got supplier personnel in here looking at everything, running tests that we used to run, balancing systems, and much more. We don't do that anymore. It's done by supplier personnel at a much lower cost than we could achieve internally."

Chrysler has been able to reassign personnel to improve activities closer to Chrysler's core business of building cars. Because PPG staff specialize in chemical systems and have the expertise and resources of their company at their disposal, they can be both more effective and efficient in performing these activities. They also perform higher value-added work from PPG's perspective. Instead of trying to sell more chemicals, they are building customer loyalty and expanding PPG's chemical services footprint within existing customers' accounts.

Customer loyalty is a critical component of PPG's long-term strategic marketing program. Though Chrysler expects continuous improvement from its suppliers, it recognizes that the constant threat of being displaced by a competitor does not produce the greatest performance from a supplier and is not in its long-term interest. "An important benefit for PPG is that they have a captive market," said Conrad. "They have our business and they know we aren't going to bring someone in here tomorrow against them."

Nevertheless, Chrysler maintains a keen eye on costs and expects suppliers to respond. As Chrysler's Ernie Schmatz explained, "We monitor costs constantly. When you pay by the unit, as you examine usage, you're examining costs. And that's the whole point."

The result is that the future of Chrysler and PPG is critically linked, with each trying to help the other improve processes and increase profitability. As PPG's Petty put it, "It's not a program that was set up to take advantage of one party or the other. It is a program to optimize the system for both parties in a partnership—either losing or gaining together. So when Chrysler gains, we gain. When Chrysler loses, we lose. That's an incentive."

THE PROBLEMS

In most corporations, there is no shortage of new programs. Few succeed and become established corporate practices. What was different about Pay-as-Painted? Petty explained:

> I came out of the steel industry and I know—there have been so many programs. They come around each year. It starts at the top with 'this is a new thing, we all need to do this," but by the time it gets down to the front-line supervision, it seems like another pie-in-the-sky project and it fails. Pay-as-Painted was a program that started that way, but it worked. You had champions of the program at each plant who made sure that things got done. Also, everyone was convinced of the savings and the efficient operation of the system.

Of course, as with most firms, the notion of relying on one supplier for a wide array of chemicals and chemical services made some people nervous. But Bob Conrad explained Chrysler's philosophy:

> We had to look at the big picture and get together with our supplier. It meant putting a lot of eggs in one basket, but we had to face the facts that only if we get our suppliers to cooperate and become part of our team can we solve some of the tough problems and stay competitive. Yeah, I may even have to pay a little more for purple paint than I did, but I'm going to save a ton of money and trouble in other ways—reduced waste costs, improved environmental performance, better relations with our neighbors.

Tim Gillies, the first on-site manager for PPG, recall a vivid memory of the difficulties in getting the program started and showing that it can work:

> A lot of responsibility for monitoring the program was given to the controller's office. At the beginning of the program—and I'll never forget it as long as I live—we were up there meeting the controller in charge of the program. He started by telling us, "I don't trust you, I don't like you, and

I don't think this is going to work." After getting over the shock, it became clear that this guy was our target. We had to do whatever it took to prove this program to him.

And, you know, he really made us better. I credit him for making us focus. He made it clear we had better have everything straight every day. Guys like him are hard to win over, but we won him over by proof, by facts. We put it in front of him, and if we were wrong, we said we were wrong. If we were right, we stood our ground and he respected that. He finally became one of the program's strongest proponents. He called me up before I left and told me how much respect he had for me and the program. He was really grateful that it had worked.

The long-term relationship between Chrysler and PPG, as well as the thorough study of the painting system by both companies, helped make the implementation of Pay-as-Painted a relatively smooth process. The immediate results convinced everyone that it was a program worth keeping.

THE FUTURE

Whether the Shared Savings program at Belvidere will expand to include even more chemicals and chemical systems is unclear. Though many chemicals, such as those used in water treatment, are not currently covered under the program, they are a much lower priority for Chrysler in terms of cost and environmental impact.

Both Chrysler and PPG intend to move the Solvent Management program to a unit pricing scheme. It's logical to harness the same incentives that have proven so successful in Pay-As-Painted. However, it is going to take time to collect the data and develop an understanding of the process sufficient to establish appropriate unit prices as well as baselines against which to measure progress. The Pay-as-Painted program is also likely to continue to expand into additional Chrysler plants. PPG's experience will prepare it to compete for these new contracts. Despite the progress and the success to date, both Chrysler and PPG believe that they have more to learn about this innovative approach to chemical supply.

19

General Motors Electro-Motive Division

Shared Savings Case History
General Motors Corporation
Electro-Motive Division
Locomotive Engine Plant
LaGrange, Illinois
and
D.A. Stuart Company

Mike Podolak could hardly believe it. Months of effort were finally paying off. A team had been studying ways to improve the central coolant systems for machining operations at the plant. After several disappointing attempts, they seemed to have found the right combination. Coolant consumption had been cut in half, yet the quality of the coolant had improved. The weekly addition of expensive biocides had ended, yet problems with bacterial growth had all but disappeared.

Mike was very pleased. These changes would allow the plant to dramatically cut the amount of chemicals they needed from their supplier. The unusual thing about this story, however, is that Mike works for the chemical supplier! And they were just as pleased to see the drop in demand for chemicals. It was an odd but increasingly common event since the plant started the Chemicals Management Program.

SUMMARY

The Electro-Motive Division (EMD) of General Motors has had a Shared Savings chemical management program with D.A. Stuart Company since 1994. Known as the Chemicals Management Program, or CMP, it covers the highest-volume chemicals used in the plant, including machining fluids (coolants), oils, cleaners, and water treatment chemicals. Stuart is paid a fixed monthly fee and, in return, provides a full array of chemical management services, from chemical ordering to assisting with waste disposal. Stuart retains ownership of the chemicals until they are used by EMD.

The plant has experienced numerous benefits from the program, including a reduction of more than 30% in chemical purchase costs. The number of coolants and cleaners used in the plant has been reduced dramatically. The supplier has successfully eliminated the need for biocides in the major coolant systems. The newer, higher-quality coolants have significantly longer life, reducing the amount of coolant consumed as well as the production downtime required for change-outs. Machine tool life has been extended. EMD has experienced improvements in health and safety conditions, a considerable reduction in required chemical storage space, easier environmental compliance, and a greatly simplified purchasing process.

THE LAGRANGE PLANT

The first locomotive was built at the LaGrange, Illinois, complex in 1932. The facility was designed to build switching locomotives. At its peak, the plant employed fifteen thousand people and produced five and a half locomotives each day. It was a highly integrated facility, producing most of the parts needed for each locomotive. But all that changed with federal deregulation of the railroad industry in the early 1980s. The wave of mergers and acquisitions dramatically reduced the number of U.S. railroad customers from twenty or thirty in the 1980s to only six today. Consolidation of assets in the industry resulted in a surplus of locomotives. Coupled with a growth in domestic and international competition, EMD experienced a sharp and prolonged decline in demand.

GM responded by restructuring its locomotive business, outsourcing some operations and shifting others to plants at other locations. Today, the LaGrange facility focuses largely on locomotive engine production, employing approximately 2,700 people. Final assembly of locomotives takes place at a Canadian plant.

Though today's plant represents only a fraction of the original complex, it is still a large industrial manufacturing and assembly operation dominated by heavy machining operations as well as assembly. The plant historically used large volumes of machining fluids, cleaners, oils, and other chemicals common to such operations. The plant produces a number of hazardous and nonhazardous waste streams as well as discharges from its wastewater treatment plant. Because of the plant's location in the Chicago area, control of airborne emissions is also a primary concern.

D.A. STUART

D.A. Stuart is a supplier of a wide range of machining fluids, oils, cleaners, and other chemicals. They currently manage a number of Shared Savings chemical management accounts.

THE EMD-STUART RELATIONSHIP

We had an industrial-size can of worms.
—PATTY GAWLE, *Materials Engineering, EMD*

The Contract

Although chemical management programs were used in some GM plants since the mid-1980s, the LaGrange plant did not enter into its program with D.A. Stuart until 1994 (see Table 19-1). The program is coordinated through a plant Chemical Management Committee consisting of representatives from many different departments. Under the program, Stuart provides chemicals and an array of chemical management services.

The current contract has two parts. The first, covering high-volume chemicals—coolants, cleaners, oils, and water treatment chemicals—employs a fixed monthly fee to reimburse Stuart for both chemicals and management services. The second part of the contract, covering an array of lower-volume chemicals—such as bulk and cylinder gases—is set up on a management fee basis. Stuart buys the chemicals from other suppliers and the purchase price of these chemicals is passed through to EMD. Stuart receives a fee for the management of the chemicals.

Table 19-1.
The CMP Contract

1. **Chemical footprint**—machining fluids, cleaners (metal washing, degreasers), oils (hydraulic and dielectric fluids, quench oils, circulating oils and greases, etc.), water treatment chemicals (power house, cooling towers, wastewater treatment, deionized water system), and miscellaneous low-volume chemicals
2. **Financial relationship**
   - Fixed fee per month for fluids, cleaners, oils, and water treatment chemicals
   - Management fee for services related to miscellaneous low-volume chemicals and filter management program
3. **Risk/Reward**—adjustment of monthly fee if production deviates by more than 10%. Mechanism for Stuart to share unusually large savings with GM–EMD.
4. **Responsibilities**—GM's UAW employees provide most of the hands-on work, including chemical changes and additions. Stuart's activities include:
   - Purchasing and inventory control
   - Monitoring and coordinating of chemical usage
   - Research and improvement of process performance
   - Product selection, testing, and analysis
   - Problem resolution support
   - Specification lab facilities
   - EHS compliance and training
   - Continuous waste minimization
   - On-site lab procedure audits
   - Tier 2 development and management
   - Data processing and clerical support
   - Interface with chemical users, particularly Production, Maintenance, and Purchasing
5. **Liabilities**—basic provisions for ownership of chemicals, prohibition of silicone-containing materials, and financial commitments. Liability associated with individual events is determined on a case-by-case basis.
6. **Performance requirements**
   - Annual 3% reductions in fixed monthly fees
   - Reductions in the number and volume of chemicals
   - Regulatory compliance assurance
   - Reductions in waste

The People

Unlike the GM Truck and Bus plant, where CMP is affiliated with Engineering and Environmental Control, EMD has placed the program with Maintenance, particularly Preventive Maintenance. Darlene Adams, a chemical buyer for the plant, explained the reason: "It's best to be close to your users. The biggest complaint purchasing departments often get is that they don't understand the user's needs. We wanted our supplier's people sitting right next to their users."

At EMD, the maintenance department uses most of the chemicals. Machining fluids, cleaners, and oils are among the highest-volume chemicals used in the plant. Maintaining the production systems that use these chemicals is the responsibility of the maintenance department.

However, many other departments also use chemicals. In recognition of this, the Chemicals Management Program at EMD operates through a Chemicals Management Committee. In addition to Stuart's on-site chemical manager, the committee includes representatives from these departments:

- Maintenance
- GM safety
- United Auto Workers (UAW) safety
- Industrial hygiene
- Environmental control
- Materials engineering
- General stores
- Purchasing
- Production
- Utilities group (boiler house, waste water treatment, etc.)

The primary purpose of the committee is to pool the expertise of its members to ensure the best chemical-related decisions. Ed Vacherlon, CMP manager for EMD, explained:

"After we selected D.A. Stuart as our chemical supplier, we sat down with the committee and said, "We are new at this. How do we do it?" The first thing we did was develop our ten-step process. Any chemical change we want to make, this is the process it has to go through. It keeps everyone informed. Everybody is aware of what is going on throughout the plant. It's not limited to new chemicals that are coming in but also existing chemicals that might be used in a different application or a different process. When a chemical change might affect a machine, even the people who work with that machine must approve the change before we can move forward.

Most of the time the process doesn't take that long. We begin early on with discussions in our regular meetings. Then it usually takes just a few days to write it up, draft a justification, then take it around to everyone to sign off. I do the final audit, make sure all the signatures are there, and that everyone has been contacted. When I sign off, it's done. We have been very happy with this process.

Dan Drozd, planned maintenance supervisor for EMD, emphasized that it is the daily work of members of the Chemical Management Committee, not the monthly committee meetings, that makes the difference. "The idea is to bring together the people who are important to a decision. Meetings are not really required. At our monthly meetings we discuss things, try to see what is coming up. But we work together daily to get decisions made."

The success of the CMP program at EMD is due largely to the success of the relationships among the people involved. Several factors contribute to this success. First, the Chemical Management Committee provides an effective structure for people to interact and work together routinely. Another factor is GM management, which has set a new tone for cooperative partnerships with chemical suppliers. A third factor, which we believe to be the most important, is the process both companies use to find the right chemical management staff for the plant. This point is important to Jim Buskus, plant safety coordinator for the UAW, and a member of the Chemicals Management Committee:

> The people that D. A. Stuart puts on site are critical. You can't come into a plant with a canned package and make it work. You have to grow it once you're there. The [Stuart] people need to do that. At this plant, there is feedback going on, discussion back and forth. It's the synergy of these two organizations' people that makes it work. Even with the right supplier, the program could fail if they used the wrong people.

To assist with the selection of the best people for the plant, EMD's Chemical Management Committee developed a supplier representative selection process that was included in the contract. According to Ed Vacherlon:

> We stated in the original bid spec the qualifications we were looking for in an on-site person and required that the candidates be interviewed by the team to discuss expectations. The entire team had to agree on the person. If that person does not work out, it is also written into our contract that Stuart will remove that person and someone else will be brought in.

After selecting the first on-site chemical manager, D.A. Stuart worked to assure an orderly transition as the EMD staff move to other positions in the facility. To train additional chemical managers that could support

EMD, Stuart implemented a rotation program where new chemical managers are rotated among many plants, including EMD, for periods of a few days to a few weeks. This way, Stuart personnel become familiar with a variety of plants, their operations, and their personnel. EMD, on the other hand, has the opportunity to work with a number of Stuart employees. When the time comes to replace or place an additional chemical manager on site, the Chemical Management Committee draws on these experiences with Stuart to select the person who best fits with the needs of the plant.

EVOLUTION OF THE RELATIONSHIP

Getting Started

As with many plants we studied, it took a crisis to initiate a positive change in EMD's chemical management strategy. Historically, chemical purchases represented only a small fraction of the cost of producing a locomotive, which can sell for as much as $1.5 million. Prior to industry deregulation in the early 1980s, the plant's greatest concern was keeping up with demand—not reducing costs. But the market decline of the 1980s made price competitiveness essential. EMD had to rethink every step and every aspect of the locomotive production process.

During this reevaluation process, it became clear that the impact of chemicals extended far beyond their purchase price. Regulatory compliance, worker health and safety issues, and concerns about the wastes generated by production were becoming increasingly important. Chemical Stores was devoting almost 10,000 square feet of storage to coolants, cleaner, and oils, including one heated outbuilding specifically built to keep certain chemicals from freezing during the winter. Chemical ordering was decentralized throughout the plant. Chemical selection was, in some cases, determined by the machine operators. Purchasing had two full-time employees dedicated to searching for the lowest-cost chemical. The plant was buying hundreds of chemicals from dozens of suppliers.

As management restructured its staff and worked to trim costs, it recognized that a chemical management program could help the plant meet its quality and cost-reduction objectives. A representative from GM's corporate CMP group was invited to the plant to make a presentation. Ed Vacherlon was asked to head the project for EMD shortly after he joined the company in the early 1990s.

In 1993, Vacherlon assembled a team representing every area of the plant affected by chemicals. The first tasks of the team was to identify chemicals to be included in the initial program and to document their baseline usage.

The team realized that the traditional approach of buying chemicals on a per-pound basis was not beneficial for the plant. "We had so-called 'chemical management' here before," commented Ed Vacherlon. "These suppliers would say they were our chemical managers when in reality they were chemical salespeople; they were just selling us product." Jim Buskus, the UAW representative on the Chemical Management Committee, interjected, "And they were very good at selling us products! Especially on the high-volume lines, they use to love to change it out, no problem! We had very few metrics to find out what was going on with our system. We relied on the salesman. That's like relying on a fox to watch the henhouse!" "That supplier had been at the plant almost ten years," noted Vacherlon. "He must have been crying when he left here."

The group then prepared a prequalification survey for interested suppliers. Ed Vacherlon recalled the team's surprise with the vendors' response to the survey: "We sent out the prequalification survey to about twenty companies. We got only eight or nine back. We were surprised; we thought we would get many more! But this made it clear which companies were serious about chemical management."

From the responses to the survey, the team selected five companies from which they requested a bid specification. They invited each supplier to spend two days in the plant. The EMD plant manager began each visit by laying out EMD's goals for the program. Then each of the chemical management team members made a presentation on the chemical needs and expectations of the areas they represented. This was followed by a tour of the plant. These visits were conducted as open discussions in which each side was supposed to learn as much as possible about the other. All suppliers were welcome to return to the plant if they had additional questions, and several of them did.

One month later, the suppliers returned to give a presentation to the team. They were required to explain why they wanted to work with EMD on chemical management and what they could bring to the plant through the chemical management relationship. Again, this was followed by open discussion of questions or concerns regarding the new potential relationship.

Technical and financial aspects of the bid were evaluated independently. Financial bids were reviewed by the purchasing department. The team reviewed only the technical specifications of the bids. From the bids and from the meetings with each of the suppliers, the team ranked the five candidates. Analyzing each of the suppliers' technical and financial specifications was not an easy process, but once completed it made the task of ranking the suppliers much simpler. Ed Vacherlon reflected on the effort involved:

> The rule for everyone at EMD was that you had to be involved in every step to make the selection. You couldn't sit out and say, "Well, I'm going to be on vacation so I won't get to see this company." We lined up the plant tours to run for two weeks straight. It wasn't easy for any of us, but that's how important we believed Chemicals Management was. In the end, all that hard work made the selection process simple. It took us maybe fifteen minutes to do our final ranking.

Using the technical rankings from the Chemicals Management Committee and the financial bids from the suppliers, it was up to the purchasing department to negotiate competitive terms with the supplier ranked first by the team. If terms could not be agreed upon, purchasing would move to the number-two supplier, and so on. D.A. Stuart was the number-one choice of the team. Though they presented the best all-around qualifications, the team was particularly impressed by their presentation, according to Jim Buskus, UAW safety coordinator:

> One reason we picked D.A. Stuart was their presentation. They acted as a team. The leadership was there, and they set the tone for the presentation. But when anything had to be answered, they relied on the specialists to do the answering. And the specialists didn't wait for the manager to answer. They picked up on the cue and they gave the answer. That was the kind of initiative we wanted in the plant. The manager is supposed to set the direction and the team is supposed to pick up the cue and get you there.
>
> In some of the other organizations, we saw a lot of salespeople whose knowledge of the process was superficial. We asked for statistical process control charts. One group brought some in, but it showed us charts of systems that were out of control. We asked, "What did you do to fix it?" They were at a loss! If it's out of control, you have wide variation. You don't know when to add chemicals and when not to add. How can you manage it? Those aren't the kind of people we want in our plant.

The First Contract

The initial three-year contract was limited to coolants, cleaner, oils, and water treatment chemicals. Stuart served as the Tier 1 supplier for these chemicals, managing contracts with Tier 2 suppliers. The monthly fee was based on historical chemical usage and costs for each of the chemical systems covered by the contract as well as cost savings that Stuart believed it could bring to the chemical management process. This amounted to an initial reduction of 30% in EMD's chemical costs. In addition, Stuart agreed to accept a 6% annual reduction in its fee.

The primary goal of the initial contract was to improve chemical processes, simplify logistics, and ensure chemical safety. A number of projects were initiated to improve coolant and cleaner usage. Chemicals and chemical suppliers were consolidated. Inventory was reduced through both the consolidation of chemicals and the use of a just-in-time (JIT) supply program from Stuart's local warehouse. Chemical safety became a key criterion in the Chemical Management Committee's ten-step approval process.

The success of the first contract was clear to everyone: both EMD and Stuart were satisfied with the relationship. At the end of the first contract period, EMD decided to renegotiate the contract with Stuart rather than to rebid. The company was pleased with the supplier's performance and wished to maintain the momentum that the program had established.

Contract Renewed

The new contract is broader in scope and more comprehensive than the original. The five-year agreement includes the original chemical footprint and fixed monthly fee plus a new set of responsibilities with a corresponding expansion of management expectations. This new pass-through program covers an increased number of low-volume chemicals to be managed, including bulk and cylinder gases. The cost of these chemicals is passed through to EMD. D.A. Stuart personnel provide logistic and management services from ordering to container return.

The new pass-through contract also includes the Filter Management program. As when chemicals were originally assessed, plant management determined that over six hundred different filters were used in the plant and purchased from over 150 filter suppliers. In one case, the same filter was purchased from six different suppliers. EMD had assigned six different part numbers and issued six purchase orders each time the same filter was ordered. Again, as with chemicals, plant management decided to consolidate all filters under one program. Under the new contract, Stuart became the Tier 1 manager for all filter purchases.

THE BENEFITS

What I needed were more solutions, not more problems.
This program brought us solutions.
—Dan Drozd, *Planned Maintenance Supervisor, EMD*

Chemical Costs

The most immediate and obvious benefit of the Chemicals Management Program was a reduction in chemical purchase costs. Compared to the baseline monthly cost of chemicals in the first half of 1994, the contract with Stuart was almost 30% lower (see Fig. 19-1). An additional 6% reduction in monthly costs was achieved for each of the next two years. Under the new contract, Stuart guarantees a 3% annual reduction in fees. Though GM would not allow actual financial figures to be published, they were obviously substantial, and savings have rapidly accumulated.

Sodium Nitrite Usage

Three large-volume parts washers were using a rust preventative (RP) containing sodium nitrite, a SARA 313 reportable chemical. The use of sodium nitrite solution had been increasing, from about nine drums per month in 1996 to more than eleven drums per month in the first half of 1998. Approximately one drum of cleaner was also used each month along with the RP. EMD was concerned about the increased use of RP and its potential impact on employee health and safety.

A team of Stuart and EMD personnel analyzed the process and were able to identify a superior cleaner-RP for the system. The new solution was less hazardous and contained no sodium nitrite or other SARA 313 material. In addition, it had a longer process life, performed better, and cost less than the previous RP!

As a result of switching to the new cleaner-RP and improving the

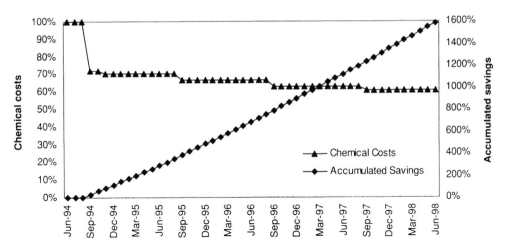

Figure 19-1. Chemical costs and accumulated savings compared to baseline monthly chemical costs, GM's Electro-Motive Division, June 1994–June 1998.

control of chemical feed rates, usage of cleaner and RP dropped from almost 12 drums per month to 2.5 drums per month, an 80% reduction in chemical volume. In addition, the new chemical cost 25% less per drum. Together, that produced an overall 85% reduction in chemical costs! Moreover, the new cleaner-RP performed much better. Rust problems related to the cleaners were practically eliminated, further reducing EMD's expenses for production downtime, scrap, and rework.

VOC Emissions

Located in the Chicago metropolitan area, the EMD plant has placed a high priority on controlling volatile organic compounds (VOCs). Clean Air Act requirements continue to tighten restrictions on VOC emissions. In their current search for remaining sources of VOCs in the plant, EMD and Stuart identified a parts washer that uses a common petroleum solvent. Though the solvent has excellent cleaning capabilities and minimizes rusting, it is a major contributor to the plant's overall VOC emissions.

EMD and Stuart personnel formed a project team to convert the process to aqueous cleaning. EMD will pay for the new washer, but Stuart is responsible for research, development, and testing of the new aqueous cleaner. Once the project is completed, the new washer should eliminate VOC emissions from the process and cut overall plant emissions in half. In addition, chemical costs should decline. Though the aqueous cleaner will cost four times as much per gallon as the original solvent, Stuart expects as much as a 95% reduction in the amount of cleaner used per year. EMD will benefit financially from the reduced VOC emission, but employee health and safety concerns will also improve with elimination of the petroleum solvent.

Coolant and Biocide Usage

EMD uses two large-volume central coolant systems for many of its machining operations. In 1994, shortly after the inauguration of the Chemicals Management Program, one of Stuart's first responsibilities was helping to solve some of the problems with these systems. One of the most common problems was bacterial growth. Over time, the coolant became rancid due to bacterial growth. The bacteria created offensive odors, produced dermatitis among workers, and caused the coolant to separate into oil and water components, thus reducing its effectiveness. In addition, EMD was experiencing periodic releases of ammonia from the coolant systems. The cause of the ammonia odor was unknown, but it was strong enough to disrupt work in the area, nearly resulting in the shutdown of operations on several occasions.

EMD's solution, at the time, was to add strong biocide each weekend to the coolant systems to control bacterial growth. Two biocides were rotated to minimize the development of resistant bacteria. The biocides were highly toxic as well as expensive, and cost EMD thousands of dollars per week.

The EMD-Stuart study team examined an array of alternatives and, in 1995, the two systems were converted to a new coolant. The new coolant performed better and lasted longer, reducing the plant's coolant consumption from over 100,000 gallons per year to about 94,000 gallons per year. However, the systems continued to have problems, so the team investigated alternatives to the common practice of controlling bacterial growth using biocides. One promising approach was the use of pH to control bacterial activity. Studies suggested that if the coolant was carefully maintained at a slightly higher pH, bacterial activity and its associated problems could be significantly reduced without the addition of biocides. Using pH rather than biocides would also allow the coolant formulation to be simplified.

In 1996, the plant stopped using biocide and instead controlled coolant pH with potassium hydroxide. In addition, they switched coolant again—this time to a simpler formulation. The outcome was dramatic. Problems associated with bacteria—dermatitis, odors, and coolant deterioration—were practically eliminated. In addition, the systems no longer released ammonia, which the team had traced to an amine byproduct of one of the biocides. Once the biocide was eliminated, so was the ammonia.

The new coolant also had a longer process life. Instead of an annual coolant change-out, the large-volume systems now require emptying only once every two and a half years. This is scheduled to perform routine maintenance on the equipment; most of the coolant is reused in the system. Fig. 19-2 presents the trend in coolant usage compared to a 1994 baseline. Elimination of the biocides, introduction of the new coolant, and a better coolant management program have cut coolant usage by more than half and improved employee health and safety in the plant.

Maintenance and downtime costs have been reduced because the coolant is removed only biannually. EMD has also saved money on the disposal of waste coolant because waste coolant haulage declined as coolant volume declined. The new coolant also provides substantially better rust prevention

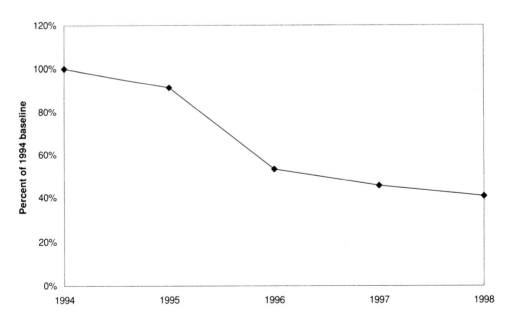

Figure 19-2. Reductions in coolant costs and usage, GM's Electro-Motive Division, 1994–1998.

than the original coolant, dramatically reducing downtime, scrap, and re-work expenses. In fact, EMD used to have a line item in the budget for rust rework in these areas. The rust rework line item is no longer needed.

Other Benefits

The total benefits generated through CMP extend far beyond savings. For example, consolidation of chemicals has provided logistic and management savings. Before CMP, the plant used more than a dozen coolants from a variety of suppliers. Now two coolants meet 99% of production needs. The same kind of consolidation was achieved with cleaners. Every chemical change was assessed through the Chemical Management Committee and the ten-step process.

This level of consolidation has produced a ripple effect on costs. Chemical logistics have been greatly simplified. Storage space for coolants, cleaners, and oils has been reduced by 90% from almost 10,000 square feet to just about 1,000 square feet. The heated oil house is no longer required, generating savings on fuel costs and building maintenance. More materials are stored in bulk tanks, dramatically reducing drum storage, handling, and disposal. Environmental, health, and safety concerns have also been reduced. Less employee training is required and it is easier to assure the safe handling and use of chemicals. Managing chemical waste is also easier. As Jack Brandush, EMD environmental coordinator, explained, "It's a lot easier for me to find good recyclers for two coolants than it is for ten or twenty."

The Committee's efforts to improve coolant systems reduced chemical consumption, but it had other benefits as well, including longer tool life and less production downtime in machining operations. Ed Vacherlon explained one example:

> One broaching operation uses three broaches. Each one is 6 feet tall. We used to get about 125 pieces through that operation before we would have to remove the broaches for sharpening. Each one takes 48 hours to sharpen. We have a group here devoted just to sharpening tools. After we put in the new coolant, we can get over 500 pieces through before sharpening! Not only does that save a lot of time in production, but the tool group benefits as well. They were going crazy trying to keep up with orders. Now they plan tool maintenance on a much more orderly basis.

Many of the cleaners used in parts cleaning equipment were in powder form. Workers were required to add a given number of scoops of powder to make the cleaner solution. Of course, what constituted a scoop for one worker was quite different than a scoop for another worker. In addition, the powder would typically fall to the bottom of the reservoir and dissolve slowly. Cleaner solution concentrations were checked only periodically by workers on the floor using chemical titration methods. Together, these activities led to widely fluctuating cleaner concentrations with widely fluctuating results. Not only did this process waste chemical but it also produced periodic quality problems, as parts were not properly cleaned or had excessive residuals of cleaner solution.

Stuart and the Chemical Management Committee solved this problem by switching to liquid cleaners with automatic feed controls. This allowed solution levels to be maintained much more precisely, eliminating the need for workers to handle the cleaning chemicals or to perform titrations. This significantly reduced overall cleaner consumption and improved product quality.

Another positive outcome of CMP is the problem-solving process associated with chemicals. This has been particularly beneficial to Ken Gertz, manager of General Stores:

> Before the Chemicals Management Program [was brought in], people on the plant floor used to come to me when they had a problem with a chemical. I would have to track down which chemical it was and who the supplier was. Then I would have to hunt down the salesman who sold it to us. It might take a couple of days just to get hold of him. He'd ask some questions and we'd have to study the problem. In the end, it could take twenty phone calls and more than a week to get the problem solved.
>
> "It's different now! Chemical problems go right to Stuart's on-site people. They spend a lot of time out on the floor, so they might even notice the problem before our people do. We used to get at least one chemical problem to deal with every day. Now we get one a month, at most!

From the perspective of the UAW, worker health and safety has been a big benefit from CMP. According to Jim Buskus, UAW safety coordinator:

> We didn't used to have much control over the chemicals that were brought in here. Someone would decide they wanted a certain chemical and that was it. Now things are different. UAW Safety is the first signature you have

to get on our ten-step approval process. We can make sure we are not bringing more hazards in the workplace. There always is a better product—one that can do the job without the risks. Let's face facts: ignorance caused most of our problems. The more we know, the fewer problems we have. That's been a real plus in this program.

Overall, the greatest value of the Chemicals Management Program has been the chemical expertise that is brought to the plant by Stuart. Ken Gertz explained, "We always have chemical expertise in the plant. It used to be if someone in the plant was away or on vacation, we couldn't get answers, we couldn't solve problems. By Murphy's Law, that's always when something would happen! But now, Stuart always keeps someone on site. If Mike is on vacation, someone else always fills in. It's been great for us."

Patty Gawle, plant materials engineer, noted how the new program has improved process engineering. "For a lot of our older equipment, process engineering specifications just didn't exist any more. The Stuart people have been working with us to rewrite those specifications. They have helped to identify the best coolants and oils and the proper maintenance procedures. It's the only way we can really control those processes."

Ed Vacherlon continued:

Our supplier has a wealth of chemical expertise and resources that stand behind their on-site people. Salespeople used to come in here with all kinds of products and technologies. We used to buy some of that stuff! Now they have to deal with Stuart, and you're not going to pull the wool over their eyes!

This is a partnership. That's how we approach the whole thing. That's how we keep focused. They share the risks and rewards of maintaining our competitive position. They help keep us on the leading edge of chemical technology, and who better? They are out there in the chemical world; they are a part of what's going on. They bring that knowledge to us.

THE PROBLEMS

We had to build some trust with them.
—MIKE PODOLAK, *On-Site Manager, D.A. Stuart*

Initially, the plan to develop a Chemicals Management Program did not have many supporters at the plant. Though EMD had experienced drastic staff reductions and severe budget cuts, most of the staff did not perceive that the plant had a "chemical problem." In fact, chemical management at the plant was much like it is at most plants. The experience of Ken Gertz, manager of General Stores, was common. "Before this program, I guess we didn't think we really had much of a problem. Looking back now, we certainly did! But I guess we just thought that was the way it was everywhere— the way it had to be."

Without a sense of urgency about the plant's chemical management problems, the new program looked like just another change being forced on the plant from the corporate offices. "We were pretty skeptical at first," commented Gertz. "It just didn't seem to make a lot of sense." According to Brandush, environmental coordinator, "It was something new. It was change, and change is always difficult. At this plant, we have been through a lot of change in recent years. It was difficult to accept until we started going through presentations with suppliers. The more presentations we went through, the more logical it sounded. We knew more. We started to realize that we could be a lot better off!"

Jim Buskus, UAW safety coordinator, had a similar experience. "In the beginning, I was kind of leery about the whole process, to tell the truth. I've seen programs come and go in this plant. I thought it was just another program from someone higher up. But I thought, "Let's see what happens." But when we started evaluating suppliers, working as a team, it started to look a lot more promising."

Of course, to many people the new program looked potentially threatening—it threatened to change their jobs, or even eliminate them. That has been a difficult problem to work through—one the plant is *still* working through. Ed Vacherlon explained:

> There were managers out there who lost some of their autonomy. In the beginning, it was a big culture change. A lot of the guys felt that they were being questioned on their abilities, how they were doing things, what products they should be using. They had the autonomy to say, "This is what I want," and it was done just like that. Today it isn't just done like that. That first year was a real tough transition. Change is difficult, whereever you go.
>
> [Plant personnel] figured that the new chemical manager came in house just to bring in [Stuart's] own products. . . . Well, we picked the supplier based on technical merit. It's not the good-old-boy system anymore.

Mike Podolak, Stuart's on-site manager, continued:

> We had to build some trust with them. The Chemical Management Committee and the ten-step process has been critical to the success of the program. It helped to build trust during the time when the plant was making the transition to the new program. Previously, the authority to make chemical-related decisions rested in the hands of relatively few people. Under the new program, these people needed to be assured that their concerns were still critical to the decision-making process. Chemical changes had to be justified by their merits. The ten-step process was a way to do this.

The purchasing department also anticipated dramatic changes with the new program. The chemicals to be covered under the new program represented 80% of the plant's chemical usage. Under the new program, Purchasing would issue a single purchase order each month for all chemicals and associated services. As it turned out, the purchasing department was downsizing and the program fit well with their attrition plans. The union, on the other hand, was confident that CMP would not result in lost jobs. UAW workers continue to perform most chemical management activities. "We haven't lost any jobs over this," commented Jim Buskus. "I think we have even gained a little." Workers perceive the Stuart on-site personnel as the chemical experts and frequently go directly to the Stuart chemical managers to get answers to their chemical questions.

Yet some attitudes remain hard to change. "It's tough to change the way you've been doing things for years," remarked Vacherlon. "When you've changed the machining tool every fifty parts, you don't like the supplier coming in and telling you it's okay now to run two hundred parts. At EMD, quality has been the number-one goal, and any practice is difficult to change when the plant personnel believe it may affect product quality. Even with the ten-step process, some people resist the change. They keep changing the tools every fifty parts!"

THE FUTURE

At the start of a new five-year contract, the future looks good for EMD's Chemicals Management Program. D.A. Stuart's responsibilities continue to expand and everyone is more confident and comfortable with the program and what it can accomplish. The Chemical Management Committee has no shortage of ideas for improving chemical management practices. In the words of Ed Vacherlon, "This thing is a gold mine!"

PART 4

PUTTING SHARED SAVINGS TO WORK

CHAPTER

20

Benefits for the Chemical User

The five manufacturers featured in the case histories and other companies that have implemented Shared Savings chemical management enjoy a wide variety of benefits and operating advantages. Following the format introduced in chapter 3, we associate these benefits and operating advantages with:

- Cost
- Quality
- Capabilities

Cost savings are the best-documented benefits from Shared Savings programs. Thus, we turn first to quality and capability benefits, then devote the remainder of the chapter to cost reductions from Shared Savings programs.

QUALITY

As noted in chapter 7, *quality* includes not only the attributes of the product but also its less tangible aspects, such as reputation and reliable supply. Shared Savings programs can enhance all these aspects of quality.

Product quality can be enhanced by improving the performance of direct chemicals—those that become part of the product. For example, the Chrysler Belvidere plant enjoyed a number of quality improvements, from a powder antichip coating to improved paint repair, as a result of the

Shared Savings partnership with its paint supplier. However, product quality is also enhanced when indirect chemicals and technologies improve the performance of production processes. Navistar and Castrol improved the quality of engine blocks and heads by eliminating incompatibility between coolants and cleaners. GM's Electro-motive Division (EMD) and D.A. Stuart improved product quality by enhancing the chemicals used in parts washers. The Ford Chicago plant, with its Tier 1 and Tier 2 suppliers, enjoyed a superior phosphate coating by removing impurities from the phosphating bath.

A plant's ability to reliably supply its customers can also be enhanced through Shared Savings programs. Most plants reported less process downtime, both scheduled and unscheduled, as a result of their Shared Savings programs. For example, GM's Truck and Bus plant reduced downtime related to sludge cleanout. Both GM's EMD facility and Navistar's engine plant reduced downtime resulting from coolant rancidity and equipment malfunction.

One of the most significant long-run benefits is an improved environmental reputation for the companies and their products. Navistar has enjoyed extensive positive media coverage of its chemical waste reductions. Ford avoided a high-profile enforcement action by dramatically reducing VOC emissions. Most plants reduced the number and volume of chemicals reported on the Toxic Release Inventory (TRI). Several of the plants received awards from state environmental agencies.

CAPABILITY

Many benefits of Shared Savings programs are not easily linked to quality improvements or cost reductions but nevertheless enhance the plant's capabilities to produce value for customers and shareholders. All of the plants we studied increased their *adaptability* by consolidating chemicals and suppliers, using a single Tier 1 chemical supplier, and simplifying chemical processes where possible. In addition, the organizational changes that accompany Shared Savings, particularly the chemical management team, accelerated the rate at which chemical-related improvements can be implemented.

In addition, all of the plants have significantly better understanding of their processes and chemical systems. They are able to track chemical usage in much greater detail. This intelligence allows them to solve long-standing problems and to plan for product and process changes in the future. Ultimately, it means better products and streamlined production operations.

The closer working relationship between chemical user and chemical supplier enhances the supplier's ability to *anticipate* the future needs of its customers. Suppliers observe firsthand the strengths and weaknesses of specific chemicals in specific applications. They are able to see downstream problems caused by certain chemical decisions. They are able to recognize the benefits of specific improvements in chemicals and chemical technology. This is not only valuable market intelligence for the supplier, it also results in improved products and services for the chemical user.

Finally, and most importantly, the Shared Savings program enhances the learning and innovation capabilities of the plant. The most profound change that occurred at each of the plants we studied was a dramatic acceleration in the rate of innovation. Both chemical user and chemical supplier bring their core competencies to the chemical management process.

REDUCED COSTS

In the previous chapters, we illustrated that improved chemical management can reduce costs for a business in two ways—first, by reducing the chemical cost iceberg (total cost of chemical ownership), and, second, by improving the performance of operations, allowing chemical as well as other production resources to be used more efficiently. Savings can also be realized in areas unrelated to the chemicals or systems covered in the contract.

Before we examine specific unrelated savings, however, it is useful to understand how the Shared Savings contract drives cost reduction.

Savings Drivers

The Shared Savings contract divides the chemical cost iceberg into two sections (Fig. 20-1). We call these two sections the *performance expectation component* (PE component) and nonperformance expectation component (non–PE component).

As we discuss in detail in chapter 23, a key element in the Shared Savings contract is the customer's performance expectations. Performance expectations usually include services that were previously the responsibilty of the customer's personnel, such as chemical ordering, inventory management, tracking, distribution, quality control and regulatory compliance

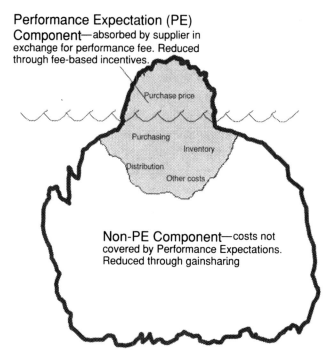

Figure 20-1. Components of the chemical cost iceberg.

activities. In meeting the performance expectations, the supplier absorbs a portion of the chemical cost iceberg, indicated by the shaded area in Fig. 20-1. We call this the performance expectation (PE) component of the iceberg. The supplier receives a performance fee for meeting the performance expectations and for the ultimate performance of the chemicals supplied.

The chemical user experiences an immediate reduction in costs for procurement, inventory, and the other services now performed by the supplier. If the performance fee paid to the supplier is smaller than this PE portion of the chemical cost iceberg assumed by the supplier through the performance expectations, then the buyer realizes an immediate net savings. The chemical customer is paying the supplier less money to meet the performance expectations than it originally cost to perform the same services internally.

In practice, we found that the performance fee may be smaller than the original chemical purchase costs alone (particularly in large companies, where a supplier can enjoy significant economies of scale). This produced a substantial immediate savings for the chemical user. The extent to which the chemical customer can negotiate a lower performance fee depends on its relative market power as well as the potential profitability of the account for the supplier. For example, an account that includes a large volume of a supplier's high-margin chemicals is more likely to provide the supplier with flexibility for fee reduction while maintaining profitability. Likewise, in plants where the supplier perceives ample opportunity for cost and chemical reductions through chemical management practices, the supplier may be more flexible in negotiating the chemical management fees.

But the savings do not stop there. The unit-price or fixed-fee strategy provides a constant financial incentive for the supplier to look for cost reduction opportunities in the PE component of the chemical cost iceberg. The more the supplier reduces the PE component costs, the greater its profit margin. For example, if the supplier improves inventory operations, which reduces operating costs, its profits increase. Most contracts include a financial mechanism that allows the customer to share these savings. Such mechanisms include fee rebates, reduced fees, and increased chemical management services provided by the supplier.

As an added benefit, many of the cost-saving activities implemented by the supplier indirectly reduce other operating costs for the customer. More specifically, cost reductions in the PE component of the chemical cost iceberg can produce corresponding reductions in the Non–PE component. The most common example is a reduction in chemical volume. Suppliers are always looking for ways to reduce the volume of chemical needed in an operation because they produce a significant reduction in operating costs. For example, substituting a new coolant with a longer working life in a machining operation reduces the amount of coolant consumed each month, thus reducing the supplier's chemical costs. It also reduces the supplier's costs for ordering, inventory, maintenance, etc. (all PE component costs). However, reducing chemical volume may indirectly reduce many chemical costs still carried by the chemical user (non–PE component costs). For example, reduced coolant volume can reduce the costs of wastewater treatment and hazardous waste disposal. In addition, lower chemical volume can reduce regulatory compliance and reporting costs.

The supplier's activities may even produce cost savings beyond the chemical cost iceberg (i.e., outside the chemical life cycle) by improving the performance of an operation. For example, the new coolant could increase the life of the machining tools. These savings can significantly reduce total operating costs by reducing labor and equipment downtime required to replace worn machining tools. In this case, savings are realized even though they are not part of the chemical cost iceberg.

Occasionally, a supplier may even identify savings opportunities in areas completely unrelated to the chemicals or systems covered by the contract. For example, at the GM Truck and Bus plant, the supplier identified an opportunity to extend the life of cloth robot covers, providing a substantial savings to GM. Again, operating cost savings can be achieved even if they do not reduce the chemical cost iceberg. Opportunities for such savings are limitless.

To promote savings beyond the PE component, many chemical users utilize *gainsharing strategies*, which allow the chemical supplier to share in customer savings it generates via recommendations or improvements outside the PE component of the iceberg. For example, a customer may have a problem with a machining process that is not chemically related. The chemical supplier may have worked with another manufacturer who had a similar problem and can assist this customer with resolving it. Gainsharing provides a financial incentive for the supplier to assist the customer with problems beyond the scope of chemical management contract. The intent of gainsharing is to create a financial incentive for chemical suppliers to make money on their expertise and experience that extends beyond chemicals. Gainsharing drives the pursuit of cost-saving innovations beyond chemicals by giving the chemical supplier a stake in the financial outcome.

Reducing the Total Cost of Chemical Ownership

A chemical supplier can offer a range of services to the chemical customer to reduce the chemical cost iceberg. The supplier can provide a broad range of services or concentrate on a given area of the chemical life cycle. In general, a company's chemical life cycle costs can be grouped into three categories:

1. Logistic
2. Environment, health, and safety (EHS)/compliance
3. Application

Logistic Benefits

Two of the most significant logistic benefits that can be obtained though a chemical management program are reduced chemical acquisition costs and chemical inventory costs. Commonly, acquisition and inventory responsibilities are assumed by the supplier as a performance expectation (the shaded portion of Fig. 20-1). As a result, the chemical customer receives these savings right at the start of the contract and every year thereafter. Once responsibility for chemical acquisition is assigned to the supplier, the chemical customer may need to issue only one purchase order per year to cover all the chemicals and associated services provided by the supplier under the contract. This can represent a significant savings, as many chemical users purchase hundreds of chemicals several times each year. Chemical management can significantly reduce purchasing paperwork. Chemical inventory may be managed by plant personnel and coordinated by supplier personnel, or the entire inventory may be managed by the supplier, thus saving the customer personnel and inventory carrying costs. Whether a company categorizes these as hard or soft savings depends on how the company assesses cost and utilizes the personnel time gained from shifting responsibilities to the supplier.

Additional benefits can be accrued through the non–PE component, the unshaded portion of the iceberg shown in Fig. 20-1, by improving inventory delivery and management. Improved chemical acquisition and inventory control can reduce or eliminate production delays or interruptions from running out of critical chemicals. It can also reduce the amount of in-plant inventory, reducing the floor space devoted to inventory and eliminating plant concerns about disposal of outdated or off-spec chemicals because the chemicals belong to the supplier.

Comments from customer personnel working in plants with Shared Savings programs help to illustrate these benefits:

- "... [M]aterials Handling could see the benefits of reduced inventory carrying costs and having less chemical in the plant to manage. Plus, they no longer had to be responsible for making sure the chemicals were here on time."
- "[Our supplier is] real good—they find ways to use up old chemicals and reduce inventory. They don't keep a lot of the chemicals in their inventory, which is good—we're not spending a lot of dollars on product and it reduces our risks."
- "To appreciate what this [Shared Savings] program has done, you have to understand what it was like here before the program. You'd see what a difference this has made. Before, we might have thirty-five or forty managers spread out over three shifts, each trying to make his

own chemical purchase decisions. We were using three or four hundred chemicals, so we had all these salesmen coming in and trying to sell each manager on their chemicals, telling each one how great their chemicals are. There was no coordination. We had half a dozen different chemicals for the same job."

- "From a chemical safety standpoint, how do you keep track of all this? From a purely logistical standpoint, how do you do ordering and where do you store all this stuff? It was terribly inefficient. We had forty or fifty salesman tripping over each other. Taking our time. And none of them getting enough business to really make it worthwhile.
- "By going to the new program, we were able to reduce duplication. In the old system, we had accumulated these stockpiles of these chemicals throughout the plant. Some of it was stuff we had stopped using. But under the new program, the chemical manager understood how we could use it all instead of having to pay someone to come and clean it up and dispose of it."

Another important logistic benefit is performance monitoring and maintenance of the chemicals during use in production. This is particularly true for coolants, cleaners, oils, and other chemicals that must be monitored and maintained over a long period. Because the supplier is responsible the performance of the chemicals, chemical monitoring and maintenance are often responsibilities shared with the customer. Direct savings accrue from the reduced number of man-hours required to perform these maintenance activities as well as the reduction or elimination of laboratory expenses. However, significant indirect savings also can accrue from the application of higher-quality fluids and from changing the fluids only when they no longer meet performance specifications. Again, plant personnel comments made this point:

- "I think people could see the advantages of this relationship. The maintenance manager knew that without the [supplier's] on-site manager, a supplier would not share responsibility for the maintenance program."
- "In the old days, it was ridiculous the way we used to pump sumps around here. We'd fill them one shift and pump them the next. Now there is someone checking this thing—there is a system or mechanism in place."

- "If you don't monitor your systems, you get microbial growth, and after a week it could be so bad you have to dump it. What [our supplier] does is make sure that each individual system is maintained."

Chemical tracking is another area in which the chemical user enjoys substantial indirect benefits. Simply stated, chemical tracking is the documentation of the chemical logistics process. Not only does it help improve chemical management, it can reduce EHS/compliance costs and improve application efficiency as well. For example, at Navistar, the chemical supplier introduced a new chemical tracking system and found a sharp increase in fluid use in one parts cleaner. Closer evaluation revealed a malfunctioning fluid controller, which was easily repaired. This reduced the wastewater treatment load as well as fluid usage.

EHS/Compliance Benefits

The chemical users we studied were extremely pleased with the EHS/compliance benefits they received from their Shared Savings programs. Supplier activities included such services as overseeing chemical shipments, providing employee training, and assisting in regulatory reporting. Comments from plant EHS managers reflect these benefits:

- "Before this program, there was no light at the end of the tunnel for my responsibilities. Now that has all changed. Certainly one of biggest benefits with this program is that we have been able to get the management help that is necessary for continuous training, continual improvement in waste reduction, and even in improving quality. We've been able to do this without raising costs, and in some instances the costs have gone down."
- "We write fewer manifests, we have less risk, because we are not running as many waste trucks out of the plant. There is less time spent filling out manifests and pumping out equipment."
- "When DEA in the metalworking fluids became a TRI material and had to be labeled as such, [our supplier] took up the challenge to develop a material to replace the fluid. [Our supplier] formulated that out for us and helped us win the Governor's award for pollution prevention. [Our supplier] worked at it and came up with a good product to replace it that is DEA-free. We have to compliment our supplier's people on their hard work."

- "We've got [our supplier's] rep here, but in addition to that, we've got all the people that are behind that rep who are here basically every day. Like the other day, questions came up about coolant that had been in the system for quite some time. [Our supplier] has a PhD toxicologist that they can call."
- "We have always made every attempt possible to reduce our level of solvents, but we're approaching it even more aggressively with the Total Solvent contract. [Our supplier] is allowing us to reduce VOCs with *much more vigor than we could have on our own.*"
- "Chemical tracking has improved environmental reporting substantially. That used to be the hardest part in doing my Form Rs."
- "Ten years ago, we would get some materials in here that didn't even have toxicology clearance. This was a concern for employees' right to know because you would be using a material that you don't necessarily have an MSDS [material safety data sheet] for. But it's different now. Now it's much easier to keep control over all the products."
- "Emission reductions have been the main improvement, from my viewpoint."

The most important EHS benefit gained through a Shared Savings strategy is chemical use reduction. As discussed previously, this is also one of the most difficult benefits to gain through a traditional chemical supply strategy, but it is usually the first area to generate savings under a chemical management program. While chemical use reduction has important logistic and application benefits, the EHS/compliance benefits can be particularly substantial. Reduced chemical usage means less chemical waste. In fact, many chemical use reduction projects we studied were initiated as waste minimization projects. Reduced chemical waste means lower waste treatment and disposal costs. It can also cut reporting requirements and compliance costs and, in some cases, discharges can be reduced to levels *below* regulatory thresholds.

At Navistar's engine plant, reducing discharges to wastewater treatment and hauling costs for spent coolant waste were major motivating factors behind the Shared Savings program during its early years. At Chrysler's Belvidere assembly plant and Ford's Chicago assembly plant, reducing the use of VOC-containing chemicals was a high priority for the respective Shared Savings programs. Much of the VOC reduction was accomplished by improving the efficiency with which these chemicals were handled and applied in the assembly process. In each case, the suppliers played a major role in technology improvements and new product development that reduced VOC emissions. Such joint efforts not only improved compliance but also saved money for the users and continue to provide cost savings into the future.

Chemical use reduction generates other tangential benefits as well. It reduces employee exposure to toxic chemicals, protecting the health of the workers. This translates into less expenditure on monitoring and control equipment, personal protective equipment, training, and recordkeeping. Less chemical in the plant reduces the likelihood of an accidental spill or release, thus reducing potential environmental liability. These factors, although not easily quantifiable, can improve employee relations by providing a safer, healthier work environment.

Other EHS benefits are achieved through utilizing the core competence of the chemical supplier. In each of the plants we studied, suppliers share the responsibility for regulatory compliance. Suppliers already had extensive corporate programs, personnel, and resources devoted to keeping current with chemical regulations. The suppliers had developed techniques and tools for maintaining their own compliance that they readily shared with their chemical users.

Chemical Application Benefits

The most important chemical application benefit, as we discussed previously, is the reduction of the volume of chemicals used in manufacturing processes. Because the supplier pays for all the chemicals in a Shared Savings relationship, volume reduction is a major priority for the supplier. All of the plants we studied had impressive examples of volume reductions implemented through a joint effort with suppliers. Suppliers were able to keep chemical use to a minimum through the appropriate application and careful monitoring of the chemicals used in each process. Rather than employing wasteful dump-and-fill strategies, they implemented repolishing strategies, computer monitoring of solution concentrations, improved chemical delivery systems that reduce spilling and evaporation, and reformulated chemicals to meet user needs with lower volume.

Beyond reducing chemical volumes, however, the most critical chemical application benefit we found was improved chemical performance in production processes. We found numerous examples where the supplier studied chemical application problems and implemented solutions that were more efficient as well as effective. In some cases, the supplier even developed a new chemical product to specifically meet the user's application needs. This cooperative, proactive performance and problem-solving relationship between the chemical supplier and the chemical user was praised many times in our interviews. It produced levels of communication, cooperation, knowledge, and understanding that were not possible before Shared Savings. Again, comments from plant personnel make this point best:

- "If there is any problem here—this metal is creating a problem with this coolant and getting the pH too high—we bring that back to the supplier and they make up some product to solve the problem. If we were doing it on our own, I don't think we would have that capability. These people are professional. They are always researching new ways to make this better."
- "What's happening now is we are overseeing all the chemical management activities and letting [our supplier] deal with the details. We don't need to specialize in water or solvents or whatever. What we do is coordinate the overall program while [our supplier] deals with the specific applications."

BENEFITS UNRELATED TO CHEMICAL MANAGEMENT

The benefits of improved chemical management are significant and have no financial or time limitations. In the long run, perhaps the greatest benefits from Shared Savings come from improvements having nothing to do with chemical management. In terms of the chemical cost iceberg (Fig. 20-1), these benefits affect costs beyond the iceberg. The chemical supplier brings knowledge, experience, and a perspective unlike that of the chemical user. That ability to see old problems in a new way can produce breakthroughs.

Many of these benefits seem to appear later in a Shared Savings relationship, once the supplier develops a working knowledge of the plant and a trust relationship between chemical user and chemical supplier is established. Unfortunately, few plants currently have systematic approach to exploring or even tracking the nonchemical benefits achieved through chemical management. However, we were able to identify several examples.

- One facility had installed a new boiler and steam line system through-out the plant. Due to a design error, the material used to construct the pipe was incompatible with the steam, and corrosion was quickly destroying the pipe system. The plant was faced with a million-dollar repair expense as well as significant production downtime. Because of the chemical supplier's extensive involvement in the plant, the on-site chemical manager learned of this problem. Supplier personnel were aware of a technology for coating the insides of the pipe with a corrosion-resistant film that would stop the corrosion process. Even though the steam process had nothing to do with the chemical manager's company, he believed that he had a responsibility to help his customer resolve the problems as cost-effectively as possible. The new process worked, and at a fraction of the repair cost.
- An auto plant was experiencing problems with the treatment of wastewater from one of its processes. On-site supplier personnel were aware of a technology that had been used successfully in the food-processing industry to solve a similar problem. The supplier conducted the equivalent of $50,000 in engineering consulting services to reengineer the solution for application in the auto plant. Though the auto company had spent years trying to solve this problem, the unique background and resources of the supplier ultimately made the difference.
- At the GM Truck and Bus facility, the supplier discovered that the plant was disposing of cotton robot covers when they became soiled. Because of the supplier's business connections with an industrial laundry company, its personnel were aware of laundry technologies that could clean the covers without damaging them. They assisted the auto plant with implementing a routine robot cover laundry service. This reduced purchasing of new robot covers more than 80%.

None of these improvements involved the chemicals or chemical systems covered by the Shared Savings contract. None involved the core chemical expertise of the supplier. Yet all of them arose because the supplier was a new set of eyes viewing old problems. In each case, had supplier personnel not been so integrally involved in plant operations, the problems would have gone unresolved.

The benefits from these types of breakthroughs can quickly equal or exceed the total cost of a Shared Savings contract. It is not surprising that many companies see tangential benefits as the ultimate benefit of Shared Savings. Gainsharing is the best way for a chemical user to promote continuous and systematic attention to potential nonchemical benefits by the chemical supplier. Through gainsharing, the chemical user agrees to share a portion of the benefits with the supplier, giving the supplier a strong incentive to find and solve problems throughout the plant. Mark Opachek, with GM's WorldWide Facilities Engineering Group, put it simply: "Gainsharing is where the *real* profit is in this program."

IMPLICATIONS FOR THE MANAGER

Cost savings from Shared Savings arise in four ways:

1. Immediate savings from shifting a number of chemical management activities (the P.E. Component of the iceberg) to the chemical supplier in exchange for the supplier's fee
2. Ongoing savings from continuous improvements in chemical management efficiency (the P.E. Component of the iceberg), allowing continuous reductions in fees or increases in services
3. Ongoing savings from reductions in other chemical-related costs (the non–P.E. Component of the iceberg) driven by gainsharing
4. Ongoing savings from non-chemical-related costs (outside of the iceberg) driven by gainsharing

Consider each of these sources of savings. How substantial could such savings be in your own plant?

As we have seen, the benefits of Shared Savings extend well beyond cost savings. Improved chemical performance can enhance product quality as well as the reputation of the company. Integrating supplier personnel into the plant workforce can increase capability, accelerating the rate of innovation and the ability of the plant to adapt to change.

Yet, in order for chemical suppliers to undertake such risky programs, there must be benefits for them as well. In the next chapter, we explore these benefits. We also explore the need to see supplier profit as an investment to be managed for maximum long-term return.

CHAPTER

21

Benefits for the Chemical Supplier

The supplier's capabilities, not its chemicals, are valued by the chemical user.

Shared Savings can give chemical suppliers an advantage in an increasingly competitive market. However, understanding the benefits of Shared Savings requires viewing the chemical supply industry from a new perspective. In this chapter, we begin with the traditional view of chemical supply and profitability, illustrating the hidden costs of servicing chemical accounts. We then demonstrate how Shared Savings can help reduce supplier costs. Finally, we demonstrate how Shared Savings suppliers are realizing a market advantage in the industry.

THE TRADITIONAL SUPPLY RELATIONSHIP

A supplier makes a profit when revenues exceed costs. In a traditional supply relationship, revenue is driven by the price of chemicals and the volume of chemicals sold. To increase revenue, a supplier must either increase chemical prices or increase the volume of chemicals sold.

To reduce costs, however, a supplier must do more than simply reduce the cost of producing chemicals. To better understand this, consider a cost

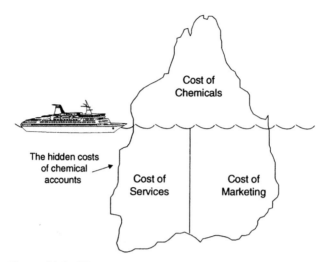

Figure 21-1. The supplier's total cost of chemical account.

iceberg from the supplier's perspective. In Fig. 21-1, the iceberg represents the supplier's total cost of a chemical account. The tip of the iceberg—the most visible component—is the cost of the chemicals themselves, including the cost of research and development, raw materials, and production. Supplier revenue minus this cost of chemicals sold produces gross margin on sales, a common measure of financial performance.

However, hidden costs are also associated with a chemical account. Two of the most significant are service costs and marketing costs. Service costs include logistic (delivery, packaging, etc.), environment, health, and safety (EHS)/compliance (material safety data sheets [MSDS], labeling, etc.), and application (assuring chemical performance) services provided to meet the performance requirements of the customer. These services may be proactive (providing the service to avoid a problem) or reactive (providing a service to resolve a problem). In either case, they can be significant.

Marketing costs are those incurred to obtain and retain an account. Obtaining an account may require years of effort by the sales staff making routine sales calls on the prospective customer. It may require the preparation of bids, coordinating chemical trials, and providing samples of the chemical products. Even after obtaining the account, it may take more than chemicals and services to retain the account. The cost of face time with customers, social activities, gifts, and other expenses can continue throughout the life of an account.

The relative size of the three cost components—chemicals, service, and marketing—can vary widely by supplier and by account. Many suppliers who compete primarily on price seek to minimize the cost of chemicals and the cost of services. The downside of this strategy is that the quality of chemicals and services can suffer, resulting in customer dissatisfaction and turnover. Increased turnover can dramatically increase marketing costs (see Fig. 21-2).

Another supplier strategy is to reduce turnover via greater investment in chemical quality and service satisfaction. This reduces marketing costs but increases chemical and service costs (see Fig. 21-3). The downside of this strategy is that the inherent customer-supplier conflict in traditional supply relationship limits the extent to which a supplier can increase chemical and service quality. Many chemical quality problems arise because the chemical is used improperly. Because supplier personnel are not integrally involved in plant operations, they have little influence of chemical management and application. Similarly, many service problems arise because of communication, cooperation, and coordination problems between supplier and customer. For example, purchasing and delivery problems often occur because chemical inventories are poorly managed within the plant (e.g., the plant uses a shake-the-drum inventory control system). Thus, pursuing a chemical quality and service satisfaction strategy can help a supplier reduce marketing costs, but the benefits are limited by its arm's-length relationship with its customer.

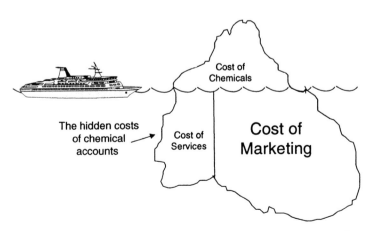

Figure 21-2. The hidden consequences of a traditional low price supply strategy.

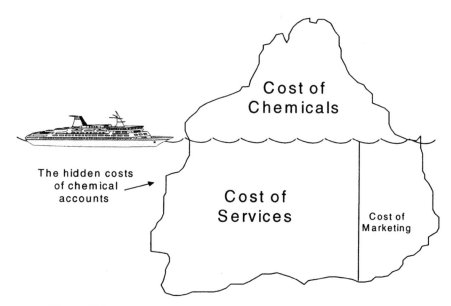

Figure 21-3. The limits of a traditional high quality supply strategy.

SHARED SAVINGS CHEMICAL MANAGEMENT

If the supplier does not sell chemicals, what does it supply?

In a Shared Savings Chemical Management program, a chemical supplier makes more money by helping the plant reduce chemical use. This is not the normal marketing behavior of a typical chemical supplier! In a Shared Savings program, it is appropriate to ask, "What is the supplier actually selling?" If not chemicals, then what business is the supplier actually in?

From one perspective, the Shared Savings supplier sells what all suppliers attempt to sell—customer *value.* The difference between suppliers in traditional sales relationships and those offering Shared Savings relationships is how they create value for their customers. The pricing strategy employed in traditional chemical supply relationships (dollars per pound) is a grossly inaccurate, even backward measure of customer value. It rewards the supplier for increasing chemical volume and creates the illusion that *chemical sales = customer value.* Suppliers naturally perceive their core business as selling chemicals.

However, as illustrated in previous chapters, true customer value is created in the customer's production operations. Chemicals can create value when they contribute to improved product quality, but chemical-driven costs serve only to *deplete* value. In Shared Savings programs, the supplier seeks to maximize the value created by chemicals and minimize chemical costs that deplete value. For all intents and purposes, suppliers no longer sell chemicals—they sell chemical expertise.

Selling chemical expertise is an entirely different business than selling chemicals. It requires different marketing strategies and sales techniques. It also requires different measures of performance and success. The supplier can no longer measure success by increased chemical sales volume or chemical sales market share. Even the term *gross margin* becomes misleading as chemical volumes decline and the chemicals are never actually sold to the customer. Instead, the supplier must shift performance focus to *net profit* and the entire cost associated with maintaining and servicing a chemical account.

REDUCING TOTAL SUPPLIER COSTS UNDER SHARED SAVINGS

Reducing Chemical Costs

Under a Shared Savings program, a supplier can increase profits by reducing the customer's chemical use. In fact reducing chemical use becomes a primary strategy for improving profit margins for the supplier. Unlike a traditional supply relationship, in which reduced chemical usage results in reduced revenue, Shared Savings programs do not link supplier revenue to chemical volume. As chemical use declines, supplier's profits increase. Suppliers consider this benefit when bidding on a Shared Savings contract. There must be sufficient opportunities to increase chemical use efficiency in order for suppliers to justify the up-front expenses when establishing a Shared Savings program, as well as any chemical discounts that might be offered in the contract bid.

For suppliers, the accounts with the most profit potential are those that offer the greatest need for improvements in chemical management practices and chemical use efficiency. One supplier's comment was typical of many we heard: "There are plants that I walk into and see that a chemical system is mistreated. There is no doubt in my mind that they are spending way too much money. I know, if they did it correctly, they don't need to spend that much money."

Reducing Service Costs

Suppliers can also decrease their service costs. Compared to chemical users, most suppliers possess greater chemical knowledge and expertise as well as greater operating economies of scale in chemical logistics, EHS/compliance, and chemical applications. In a Shared Savings relationship, suppliers manage the chemical user's inventory. Most suppliers find that managing their customer's inventory gives them an opportunity to synchronize chemical production and delivery more effectively with their customer's chemical needs. Chemicals can be matched more effectively to the customer's specific applications in the plant, reducing time-consuming and costly performance problems associated with misapplication of chemicals. With the knowledge, the supplier can also schedule and coordinate ordering and delivery more efficiently than the plant's personnel can. Emergency or after-hours deliveries can be dramatically reduced or even eliminated. Consolidating and simplifying the inventory of chemicals in the plant can reduce EHS service costs for the customer as well.

Many other sources of chemical efficiency are less obvious. For example, suppliers have the resources and expertise to identify chemical problems before they get out of control—and costly. This was noted by Bob Hendershott, the Castrol representative at the Navistar engine plant:

> Being at Navistar on a regular basis allows me to pull samples and do testing on a regular schedule. I identify problems as they occur, so problems do not continue very long before I intervene. Therefore, problems don't continue for two weeks until someone calls and says, "Castrol, your product isn't working. We got two weeks of bad parts because of you guys."

Reducing Marketing Costs

Surprisingly, efforts to reduce chemical and service costs can result in a decrease in marketing costs for the supplier as well. Because reduced chemical and service costs can increase customer value, they can create greater customer loyalty. Once the volume conflict is removed from the relationship and supplier personnel are integrated into plant operations, the increased value generated through the chemical management program produces higher profits for both chemical user and chemical supplier.

Shared Savings relationships help to reinforce long-term customer loyalty. Once in the plant, the supplier has the opportunity to develop a detailed understanding of the customer's needs, which would be difficult for a competing supplier to duplicate. This gives the incumbent supplier the ability to develop products and services that can uniquely meet the identified needs of the customer, creating additional value in the relationship. For a customer to switch suppliers would require guiding the new supplier through the learning process all over again. A Shared Savings relationship creates customer loyalty because it facilitates the ability of the supplier to provide superior performance and rewards the supplier accordingly. In addition, it gives the incumbent supplier a competitive edge over competing suppliers, because another supplier would have a hard time duplicating its performance. Bob Conrad, at Chrysler, said it best:

> [Our supplier] knows better than we do what the chemical capabilities are. We know better than anyone what our needs are. When we get together, we combine that knowledge. It's a lot better than someone coming in off the street and saying, "I can solve your problem." They don't even understand our problems! With this program, we've got someone who's not just coming in the door to sell us things. They are here to work with us to solve the problems that we have.

The customer loyalty achieved through a Shared Savings relationship often allows renegotiation, as opposed to rebidding, when renewing contracts. The bidding process is costly in terms of time and effort for both parties in a contract. The renegotiation approach is a natural progression of a Shared Savings program, as both the chemical supplier and user are committed in achieving the best output at the least cost. It is logical for both parties to utilize a renegotiation strategy in place of a rebidding strategy.

Overall, the suppliers we interviewed considered customer loyalty one of the major advantages of a Shared Savings supply strategy. We include a few of their comments:

- This will guarantee us business over the long run. It may not seem as profitable as it was, but loyalty is there. The customer says, "You have helped us out. You don't make as much as you did before, but we are going to keep you here. Instead of a sales site, which might last one to three years, you're going to be here as long as we are here.
- It definitely benefits [us] because it locks us in with [our customer] and opens the door to expand us into other areas of the business.

Or, as one chemical customer put it, "[T]hey have a captured market. They have our market, we aren't going to compete, we aren't going to send anyone in here against them."

MARKET ADVANTAGE

As we show, suppliers offering a Shared Savings program are no longer in the chemical supply business. Instead, they provide chemical expertise. This is a logical, strategic response to what many see as the inevitable contraction of the traditional chemical industry. Greater environmental, health, and safety concerns have raised the awareness among chemical users that reduced chemical use is simply good business. Chemical users are seeking ways to reduce or eliminate the chemicals they use. They are seeking to consolidate both their chemical inventories and their chemical suppliers.

The trend in the auto industry is consistent with this view. Just a decade ago, GM, Ford, and Chrysler used hundreds of chemical suppliers, with most chemicals sourced by multiple manufacturers. However, the number of chemical suppliers that these automakers work with today—even in a traditional supply relationship—has dropped dramatically. For Shared Savings programs, the automakers generally work with only ten to twenty companies worldwide as potential Tier 1 suppliers.

This trend is already having an impact on the sector of the U.S. chemical industry that supplies the auto industry. A large segment of the industry continues to supply chemicals through dollars-per-pound or dollars-per-pound-plus-services relationships. But, increasingly, they serve as Tier 2 or Tier 3 suppliers, selling to a plant's Tier 1 chemical supplier, not directly to the plant itself. Only a limited number of suppliers are pursuing a new market identity as chemical management service companies—to gain a share of this new and growing market. The shakeout of the chemical industry is well underway.

If this approach spreads beyond automaking to other industries (as we believe it has), it will widen the gap between the traditional chemical market and the new chemical management service market. Within the traditional chemical market, the number of suppliers offering only chemical sales will decline. Profit margins will contract as competition intensifies and more accounts will be managed by Tier 1 chemical suppliers, who understand the chemical marketplace and work continuously to provide higher-value chemical products and services to the chemical users who employ them.

Companies able to make the transition from chemical supplier to Shared Savings provider will find a profitable, growing market. The companies that make the transition sooner will gain the greatest market share and reap the greatest benefits in this emerging market. Shared Savings suppliers noted that their experience with Shared Savings programs gives them an advantage over most of their competitors in winning additional business from current customers as well as generating new business opportunities with chemical users who wish to begin such programs.

Several factors contribute to this competitive advantage. First, Shared Savings suppliers have the opportunity to demonstrate their capabilities to existing customers, giving them a first-strike opportunity to win additional business within a plant as well as at other plants operated by the same customer. Second, as a supplier acquires knowledge, experience, and a documented successful track record with Shared Savings, they are more attractive to customers who are implementing a Shared Savings program for the first time. Third, because supplier staff are involved in day-to-day plant operations, the incumbent supplier possesses a better understanding of the needs of the company and the industry. Suppliers are able to profit from this knowledge by developing new products to meet the needs of their current customers and field testing the new products in their customers' plants. As one supplier put it:

> One of the advantages to this system is that it is our best market research tool. Because we become so intimately involved with our customers, we can get market research information faster than our competitors can. A lot of our product testing and evaluation occurs at our chemical management accounts, and we generate better product ideas by knowing the customer so well.

Another supplier's comment sums it up: "With chemical management, you can expand business. People in the plants even bring you business. You can be more entrepreneurial. But they are not just coming to us for our chemicals, it's more and more for our expertise."

IMPLICATIONS FOR THE MANAGER

Traditional chemical supply programs rely largely on increasing chemical volume in order to provide growth for suppliers. Attempts to reduce supplier costs are limited by customer turnover (creating added marketing costs) and by customer conflict (minimizing efficiency improvements). With chemical users attempting to reduce their use of chemicals, this can be a losing strategy.

Shared Savings, in contrast, markets chemical expertise, not chemicals. It turns declining chemical demand into profit for both chemical user and chemical supplier. Improved customer loyalty reduces marketing costs. Cooperation between chemical user and chemical supplier improves efficiency, further reducing chemical and service costs.

Yet, Shared Savings can be a risky venture for the supplier. It requires an up-front investment of time and resources for a new account. Revenues come from turning chemical expertise into improved chemical performance, not from simply delivering chemicals to the loading dock. To obtain the benefits of Shared Savings, chemical users must reward suppliers with sufficient profit to offset their risks. Shared Savings is a long-term investment in the benefits to both parties.

CHAPTER

22

Implementing Shared Savings

Some companies may be ready to move from their current supply relationship directly into a Shared Savings Chemical Management relationship. Most, however, prefer to make the change in a series of smaller steps, continually reviewing, assessing, and revising aspects of the supply relationship as it evolves. For this reason, we begin the chapter with a discussion of how companies can prepare for a Shared Savings relationship by moving up what we refer to as the *hierarchy of supply relationships*. Once a decision is made to implement a Shared Savings relationship, most companies start small, gradually expanding as mutual trust and comfort with the supplier grows. Thus, the second section of this chapter discusses the process of implementing a new Shared Savings program. Finally, we end with the evolution and maturity of a Shared Savings relationship. (For a detailed manual on implementation of chemical management programs the reader is referred to the Chemical Strategies Partnership of San Francisco: www. chemicalstrategies.org.)

THE ROAD TO SHARED SAVINGS

Companies can prepare for a Shared Savings relationship by moving up the hierarchy of supply relationships.

Many companies wish to approach Shared Savings incrementally, taking time to prepare both themselves and their suppliers for a relationship that

Table 22-1.
Hierarchy of Chemical Supply Relationships

1. **Transactional** (traditional)—"Value is in the chemical," dollars-per-pound fee structure, minimal services, and threat used to solve problems
2. **Services**—"Value is in the chemical and service," dollars-per-pound plus services fee structure, focus on external logistic and EHS/compliance services but wider array of services possible, threat typically used to solve problems but limited joint problem solving possible
3. **Limited chemical management**—"Value is in chemical management," dollars-per-pound plus services or management fee payment structure, services ranging from a focus on internal logistics to a full array, joint problem solving
4. **Shared Savings chemical management**—"Value is in performance and continuous improvement," fixed fee/unit price/gainsharing fee structures, full range of services, cross-functional integration possible, incentives used to avoid problems, joint problem solving

generates greater product value for the ultimate customer. In chapter 5, we described four types of chemical supply relationships, structured as a hierarchy or pyramid. These are summarized in Table 22-1. Companies may find it easier to move up the hierarchy one step at a time, changing fee structures and adding supplier services.

Many companies also attempt to consolidate suppliers. This can be done concurrently with moving through the hierarchy. Beginning with the use of multiple suppliers for some chemicals, companies can move to a single source per chemical. Many chemicals fall naturally into groupings, such as solvents, water treatment, fluids and oils, etc. A logical next step is to select a single supplier in each group to act as a Tier 1 supplier—either supplying its own chemicals or purchasing selected chemicals from Tier 2 suppliers. Finally, if desired, a company could move to a single Tier 1 supplier for an entire plant, or even all the plants in the company.

Fig. 22-1 provides a strategic matrix of supply relationships and supplier consolidation. It can be used to develop a plan of action for a company to move from its current chemical supply situation to whichever supply strategy it desires.

For companies that currently share a transactional supply relationship with their chemical suppliers, an appropriate first step is to add key services from those suppliers with strong service capabilities. These services should be selected on the basis of producing the greatest value-added benefits

Buyer-Supplier Relationship	Multisourcing	One source per chemical	One Tier 1 per chemical group	One Tier 1 for all chemicals
Transactional	*The Past*			
Services				
Limited Chemical Management				
Shared Savings Chemical Management				*The Future?*

Figure 22-1. Array of options for supply relationship and supplier consolidation (adapted from Bierma and Waterstraat 1996).

for the chemical user. From our experience, this service selection process generally leads to supplier consolidation as trust with a primary supplier increases over time and with experience.

Companies with well-established service relationships should encourage their suppliers to accept broader chemical management responsibilities under a management fee arrangement. This could involve the creation of a Tier 1 supply position over companies (which would become Tier 2 suppliers) in each of the chemical groups.

Fig. 22-2 and 22-3 illustrate the levels of service used by General Motors and Navistar International, respectively. Though both companies refer to all service levels as *chemical management,* the lower levels of the pyramid are better characterized as traditional and service chemical supply programs. Many companies choose to begin with limited services and expand to include additional services as they become more comfortable with chemical management and as their trust in the supplier's performance capabilities increases *and* the supplier's trust in the company's cooperation increases.

Many of the service performance responsibilities related to chemicals are transferred from plant personnel to the supplier's personnel during the implementation process. (Notice that, with Navistar's program [Fig. 22-3], service responsibilities are shifted from Navistar to the supplier.) In some cases, responsibility for a service is transferred from other outside contractors to

CHEMICAL SERVICES	LEVEL I	LEVEL II	LEVEL III	LEVEL IV	LEVEL V
Chemicals					
Consultation					
Off-site Checks					
On-site Manpower					
On-site Checks					
Off-site Support					
Inventory Management					
On-site Management Team					
On-site Checks					
Complete Chemical Process Management					
R&D Programs					
Container Management					
Second Tier Development and Management					
Product/Process Engineering and Management					
Predictive Maintenance and Development					
Chemical Maintenance Scheduling					
Safety-Health Studies					
Environmental Studies					
Assist in Environmental Reporting					
Training					
Total System Analysis and Technical Mgt.					
Problem Solving					
Usage Tracking					
Cost Reduction Program					
Coordinate OEMs					
Waste Treatment Involvement					

Figure 22-2. Chemical Management Program Involvement Ladder, General Motors Corporation (adapted from Mishra 1997a). Shading indicates services provided by supplier; cross-hatched area indicates limited services.

the supplier, thus providing immediate savings for the user. In other cases, the services provided might be new and suppliers implement them for the first time.

Finally, once the chemical user and chemical suppliers have sufficient knowledge and experience with working closely and sharing chemical management responsibilities, it is time to change to a Shared Savings relationship. It is even possible, at this point, to consolidate all chemical management services under a single Tier 1 supplier to coordinate the management of chemicals from several other suppliers, who become Tier 2 suppliers.

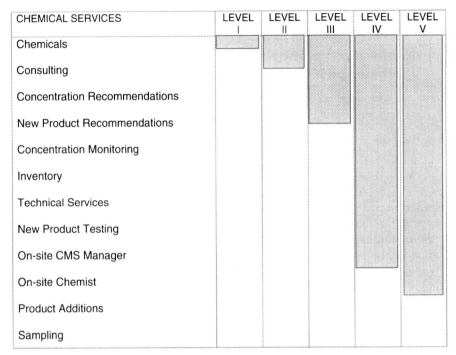

CHEMICAL SERVICES	LEVEL I	LEVEL II	LEVEL III	LEVEL IV	LEVEL V
Chemicals					
Consulting					
Concentration Recommendations					
New Product Recommendations					
Concentration Monitoring					
Inventory					
Technical Services					
New Product Testing					
On-site CMS Manager					
On-site Chemist					
Product Additions					
Sampling					

Figure 22-3. Service responsibilities in the five levels of chemical management service (CMS) programs, Navistar International, Inc. (adapted from Bernath 1997). Shading indicates services provided by supplier.

STARTING A SHARED SAVINGS PROGRAM

There is no "right way" to implement a Shared Savings program. Most companies start small and expand their program over time as they gain experience and develop a level of comfort with the concept. As an example of Shared Savings implementation, we include General Motors' approach to developing, implementing, and continually improving its chemical management program (see Table 22-2). We grouped the steps into several categories. The order of the steps in Table 22–2 is not necessarily fixed or linear, and several steps may be implemented simultaneously. In this section, we highlight a few important points about starting a new program.

Preparing Plant Personnel

As discussed in chapter 10, Shared Savings relationships are best visualized as a cycle (see Fig. 10-3) that begins with a desire to align the financial

Table 22-2.
Basic Elements in Developing, Implementing, and
Continually Improving a Chemical Management Program
(based on Mishra 1997a, and Williams et al. 1995)

Preparing plant personnel	• Concept introduction
	• Presentation to plant management
	• Management buy-in
	• Interdisciplinary team formation
	• Team buy-in
	• Team champion selection
	• Team leader selection
Footprint	• Program footprint development
Baselines	• Costbook development
Selecting suppliers and establishing terms	• Qualified suppliers list development
	• Preliminary interview with suppliers
	• Specifications development
	• Pre-bid meetings with suppliers
	• Bid clarifications
	• Technical proposal analysis
	• Financial proposal analysis
	• Supplier interviews
	• Supplier selection
	• Sourcing decision making
Implementation	• Program implementation planning
	• Program promotion/communication (in-plant)
Program growth	• Program maintenance
	• Program reviews
	• Program expansion
	• Program renewals
	• Supplier development
	• Process improvements
	• Process optimization
	• Program information and data management
	• Pollution prevention plans
	• Waste minimization plans

interests of chemical supplier and chemical user to increase product value for the ultimate consumer. The financial relationship between the supplier and the user is structured to create incentives that align their respective financial interests. This promotes joint activities that improve product quality while reducing costs, benefiting both companies and reinforcing their mutual financial interests.

A Shared Savings chemical management program cannot be implemented without a core group of individuals who understand and accept the principle of mutual financial interests and share the mindset of ultimate consumer value. Ideally, this core group includes top management as well as managers from key departments such as Production, Purchasing, and Maintenance. However, it may be possible to begin a small Shared Savings program in any area where the personnel are committed to it.

Personnel are all of the plants we studied stressed the value of cross-functional teams and a plant champion. The teams ideally include representatives of all areas of the plant that are affected, including Production, Production Control, Purchasing, EHS, Maintenance, and hourly workers. Table 22-3 illustrates the typical makeup of teams in Chrysler and GM programs. Chrysler utilizes both a corporate and plant-level teams to monitor the program's performance and progress.

Even before suppliers are selected, the plant team has a valuable role to play. First, it helps to complete the preliminary data collection and analysis work leading up to a contract. Second, the team helps personnel in all targeted areas of the plant to understand the concept and arrive at a consensus on what they want from the program. Taking the time necessary to

Table 22-3.
Composition of Chemical Management Teams
at Chrysler and GM

Chrysler Plant Team—Pay-as-Painted (Schmatz 1997)
- Paint Center management
- Paint supervision
- Maintenance supervision
- Quality Control
- Budget
- Production Control
- Hourly workers
- Supplier service reps

Chrysler Corporate Team—Pay-as-Painted (Schmatz 1997)
- Exterior Purchasing buyer
- Paint and Energy management
- Methods/Cost analyst
- Supplier sales executive
- Supplier program coordinator

GM Plant Team (typical) (Mishra 1997a)
- Manufacturing engineering
- Materials engineering
- Purchasing
- Logistics
- Financial
- Environmental engineering
- Health/Safety
- Plant maintenance
- Production
- Facilities engineering
- Power house/wastewater/cooling towers

communicate and resolve concerns during the planning process is far less time-consuming and costly than addressing concerns and problems once the program has been implemented.

We found that the team requires the leadership of a program champion. A program champion need not be a member of upper management but should be someone with enough understanding, knowledge, visibility, and respect to be influential throughout the plant. We found that champions can come from almost any area of the plant, including Purchasing, Production, EHS, and Maintenance. The champion's role is to keep the implementation and development process moving ahead as well as to motivate others despite the daily demands of production.

One plant we visited had failed to include Manufacturing and Engineering personnel in their planning team. As a result, the plant never got buy-in from these critical groups associated with product design and assembly. The trial program failed. The company had to pull back and establish a more comprehensive planning team to develop a new implementation plan—which *was* successful

Chemical Footprint: Start Small

To start the Shared Savings cycle in motion, it is best to start small. At the Navistar Engine plant, one of seventeen coolant sumps was selected for demonstration. The coolant system had the worst performance problems in the plant. The implementation team, as well as the supplier, figured that if the program could work there, it could work anywhere. In time, the Navistar program grew to include all coolants and cleaners in the plant. At the GM Truck and Bus plant, the program was limited to water treatment and paint detackification systems. Ford's program at the Chicago plant focused on paint detackification. Chrysler implemented the initial program at the Belvidere plant, addressing only the phosphating and paint detackification systems. One supplier recommends starting with chemical processes having significant potential for both cost reduction and product quality improvement.

A number of reasons support the start-small strategy. First, Shared Savings is a very different type of supply relationship than the usual. The transition from a transactional supply strategy to the principle of aligning supplier interests, sharing responsibilities, and sharing resulting benefits is difficult for many people to make in concept, let alone in practice. Most likely, pockets of employees are ready to make the change, while others perceive it as a threat. Initial success in limited applications can help convince management as well as production personnel of the potential value of adopting the new program. Also, Shared Savings requires realignment of responsibilities, procedures, and information systems. It takes time for both the supplier and the user to implement the new practices, adapt to the changes, and debug the systems.

Finally, each side must earn the trust and respect of the other. This alone takes time. Limiting the size of the initial contract limits the risk, yet provides each company with the opportunity to learn from mistakes with limited operational impact. Once the concept is successful in a small application, it is more readily accepted throughout the plant.

Developing Baselines

Performance and production baselines are essential for a successful implementation of a Shared Savings program. Sometimes referred to as the *costbook*, baselines document chemical use and costs at the beginning of the contract (see chapter 14). Both chemical user and chemical supplier require these data to determine appropriate performance expectations and performance fees for the contract. If performance expectations cover product quality and process performance, baseline data are needed to support them. The baseline data are used to measure improvement. Inaccurate baseline data dramatically increase the risks for both sides. Interviewees related a number of stories about early Shared Savings efforts where poor baseline data produced significant losses for one side of the relationship and led, in one case, to a complete renegotiation of the contract and, in another, termination of the contract.

The specific baseline data collected vary by plant and process and depend on the focus of the program to be implemented. However, there should be a link between the baseline data, the fee mechanisms specified in the contract, and the metrics used to evaluate the program (see chapter 23). This is because supplier revenue is linked to meeting and improving on performance targets. This can only be done by measuring baseline conditions and then repeating the measurements at regular intervals.

Performance baselines are often developed by the plant team prior to the contract. Information on production, chemical use, chemical cost, and other key cost or performance data are the minimum data required. At least twelve months of historical data are typically used, but longer periods are helpful, particularly if there is substantial variability in production.

However, it is not always necessary to have fully developed baselines before the implementation of a Shared Savings relationship. Ideally, the chemical user has maintained years of thorough records on chemical usage and costs for its processes but, unfortunately, this is rarely the case. Plants sometimes involve suppliers in collecting and analyzing data to develop baselines, even before a final contract. For example, at the early stages of implementing the Chemicals Management Program, the GM Truck and Bus plant invited suppliers to study various processes, establish baselines, and propose efficiency improvements. Using this approach, the plant not only received valuable baseline data but was also able to evaluate the abilities of prospective suppliers.

Still other plants involve suppliers in the process of developing baselines *after* an initial contract. This approach is common at Chrysler, whose general plan for implementing programs requires a trial period of working with the supplier. During this trial period, detailed baselines are defined and performance targets are established. Once the trial period is over and the supplier selected, the program implements unit pricing. By starting with a small chemical footprint, baselines can be developed sequentially. The Chrysler Belvidere assembly plant provides a good example. The Shared Savings contract using unit pricing was first established for selected body coating operations such as e-coat and phosphating. The processes were well understood and the baselines were relatively easy to establish. More traditional pricing strategies were used for paint supply and solvents. However, after three years of collecting and studying paint shop data, unit pricing was established for paints as well. The team then initiated a multiyear study of solvents with the intent of moving to unit pricing.

In short, while baselines are critical to a successful Shared Savings program, they should not be seen as a barrier. By starting with simple chemical systems and including the supplier in the development process, a company can easily move the Shared Savings strategy from one system to another as the program expands.

SELECTING SUPPLIERS

Select a supplier that has a reputation and position in its industry that you would like to have in industry.

The criteria for supplier selection in chemical management is different than with more traditional supply strategies. The primary value of a supplier lies in its chemical management capabilities, not in its ability to deliver chemicals at the lowest cost. A Shared Savings buyer wants a supplier who will continuously work to reduce the total cost of chemicals and to increase chemical performance now and in the future.

Selecting the right supplier is critical. Here is a list of suggestions compiled from a number of sources, particularly Dr. P.N. Mishra of General Motors (1997a, 1998), with Shared Savings experience. Look for suppliers with:

- Chemical management experience
- A commitment to Shared Savings principles (see chapter 6)
- Outstanding on-site staff
- Second-tier management skills
- Project management skills
- Budget management skills
- Process management skills, including statistical process control (SPC)
- Technical services
- Product technology
- Information management capabilities
- Customers in a variety of industries
- Relevant research and development (R&D) programs
- Career development paths for chemical management staff
- Financial and operational risks similar to those of the buyer's firm

This last criterion is important for matching buyer and supplier to avoid a lopsided relationship. If the buyer is a large multinational corporation, it is usually best to select a large multinational supplier. Smaller regional firms are best with smaller regional suppliers. A lopsided relationship is established when one company has a significantly greater financial commitment and operational risk than the other. If a large corporation asks a supplier to take on the responsibility of the chemical management program, it wants a supplier with the deep pockets that ensure it can address unforeseen financial or operational commitments without risking financial failure. Unforeseen operational mistakes could have serious implications for the buyer as well as the supplier. Conversely, a small manufacturer may not find a large supplier sufficiently committed to the success of the program because the manufacturer represents an insignificant proportion of the supplier's total revenue.

There are exceptions to the rule, of course. Individual plants belonging to a major corporation may match well with regional suppliers. Many smaller firms may not be able to find suppliers of comparable size with chemical management expertise to address their needs. These companies may need to look for larger suppliers with programs specifically targeted to smaller companies.

In addition, most companies prefer to find a supplier whose core chemical business complements a significant proportion of the chemical needs at the plant. For example, a plant beginning a chemical management program covering water treatment chemicals, coolants, cleaners, and solvents could consider suppliers of any of these chemicals but not, for example, a company supplying paints only. The reasons for this are two. First, most buyers are seeking application expertise from a supplier. A paint supplier does not possess application expertise for these other chemicals. Second, a program should be able to reduce cost if the supplier uses a majority of its own chemicals. If a Tier 1 supplier must buy chemicals from Tier 2 suppliers, it must pay prices that include the Tier 2 supplier's profit. If the supplier applied its own chemicals, its costs are limited to those of producing and delivering the chemicals.

However, some companies have established relationships with suppliers that do not manufacture chemicals at all. They selected suppliers who specialize in chemical management services rather than chemicals. An advantage to selecting a chemical management services supplier is that the company does not have an incentive to replace a plant's existing chemicals with its own. A disadvantage is that a chemical management service supplier may not possess the same level of technical resources or technical expertise as a supplier that manufactures the chemicals. This can be a critical factor when a user requires expertise and resources to resolve specific chemical application problems in the plant. Many companies that use chemical management service companies have contracts focusing on logistic services rather than application services.

The supplier selection process varies among chemical users. The GM approach is similar, in many ways, to making typical vendor selection decisions by requiring proposals from all interested suppliers. However, Ford established many supplier relationships at the corporate level. Plants and suppliers are matched at the plant level for implementation. At Navistar, the Shared Savings relationship began when one supplier stepped forward and proposed the program as a potential solution to a recurring coolant problem. All of these approaches resulted in long-term, profitable relationships for the respective companies.

Most companies request proposals from an array of suppliers. The steps followed by General Motors offer guidance and structure for the selection process (see Table 22-2). GM requests separate technical and financial proposals from interested suppliers. The technical proposals are reviewed first and a short list of technically qualified candidates is created (Mishra 1997a). The financial proposals of the companies on the short list are reviewed in detail. The logic supporting this selection process is simple: GM focuses on selecting the suppliers with the best technical qualifications first, then from among these the firm offering the best financial terms. In GM's experience, it is critical that a supplier's financial terms do not override its technical qualifications.

At least one company takes the selection process a step further. The company selection team evaluates the companies in the supply market and determines which they believe best meets the needs of the plant. The company then negotiates with the supplier to define acceptable financial terms. In other words, the selection team identifies the best supplier in the market—then it is simply a matter of negotiating financial terms. This approach makes sense when the supplier's profit is small in comparison to the total cost of chemical ownership and the performance leverage generated by those chemicals (and this is often the case—see chapter 24 on pricing).

Establishing Terms

The terms of a Shared Savings contract must satisfy the needs and concerns of both parties. However, these terms should be flexible and should be expected to evolve and change over time. It is best to begin with relatively simple contract terms until both the supplier and the user better understand the new relationship.

For example, some companies set an initial fixed fee equal to the user's chemical purchase costs for the previous year. This strategy maintains the user's costs at a fixed level for a year while it provides the supplier with time to develop an operating database and an understanding of the user's chemical processes. A supplier's chemical responsibilities may initially be limited to a small chemical footprint. In addition, the duration of the first contract may be limited to one year. Most companies progress to gainsharing programs only after several years of Shared Savings experience with a given supplier. It cannot be overemphasized that the key is to start simple and start small. In the long term, there is no limit to the profit potential of a successful Shared Savings program.

An outline of key contract terms is presented in Table 22-4. Additional details are provided in chapter 23, "The Shared Savings Contract." The most important provisions in the contract cover performance expectations and performance fees. These clearly define what the supplier is expected to provide and how it will be compensated.

Selecting the services to include in the contract is an important step in establishing contract terms. The range of services that could be included in a chemical management program is virtually unlimited. The combination of services varies by plant and over time, usually in the form of expanded services provide by the supplier. Ideally, a Tier 1 supplier is selected on the premise that more extensive services can be provided as the program expands and as the needs of the plant change.

Table 22-4.
Key Terms in a Shared Savings
Chemical Management Contract

Performance
- Chemical footprint
- Staffing and support
- Performance expectations
- Product quality
- Process performance
- Services
- Chemical characteristics
- Miscellaneous
- Performance fees

Continuous Improvement

Miscellaneous Provisions
- Liability
- Resolving problems
- Length and renewal of contract
- Termination
- Ramp-up schedule

Table 22-5.
Example Services in Chemical Management Contracts

Logistic
- Determine the need to replenish chemicals.
- Order chemicals and ensure timely delivery.
- Assure compliance with all transportation regulations.
- Manage chemical inventory.
- Take inventory daily or weekly, as appropriate.
- Distribute chemicals as needed to users.
- Collect used chemicals, as needed, from users.
- Monitor chemical waste.
- Report on chemical inventory, usage, and waste monthly.
- Develop and maintain a chemical tracking database system.

EHS/Compliance
- Assure compliance with all chemical labeling and packaging regulations.
- Conduct required chemical-related employee training.
- Advise plant personnel on safe use of chemicals.
- Advise plant personnel on proper personal protective equipment.
- Obtain all MSDSs.
- Develop and maintain MSDS database.
- Coordinate plant chemical approval program.
- Assist in plant waste minimization program.
- Assist in VOC emission reduction program.
- Assist in discharge reduction program.
- Assist in permit application process and in completing required regulatory reports.
- Track water usage in key processes.
- Assist in water use reduction program.
- Operate or assist in control of wastewater treatment system.
- Perform laboratory testing for wastewater treatment system.
- Assure proper recycling or disposal of waste.

Application
- Study and document chemical usage.
- Evaluate the performance of chemicals.
- Maintain supplier-owned equipment.
- Test and maintain chemical quality.
- Develop, operate, and maintain on-site laboratory.
- Perform R&D with respect to solving chemical-related problems.
- Recommend appropriate improvements in chemicals and technology.
- Train plant personnel on proper use of chemicals.
- Assist in developing or improving maintenance program.
- Perform maintenance program.
- Recommend appropriate chemical cleaning and reuse technology.
- Recommend ways to reduce the number and volume of chemicals used.
- Assist in process mapping and cost assessment.
- Assist in optimizing processes.
- Participate in new product design.

Table 22-5. (*Continued*)

Program Coordination
- Develop overall chemical control plan.
- Meet daily with key personnel (production, maintenance, etc.).
- Meet as scheduled with chemical management team.
- Provide minutes of key meetings.
- Assist in selecting lower-tier suppliers.
- Report on performance quarterly (savings, chemical use, waste, other).

Table 22-5 is a list of possible services that can be included in chemical management contracts. The services are grouped into logistic, EHS/compliance, and application categories. Another category of services, program coordination, is included in large programs for overall management services.

The plant's chemical management team has the responsibility to define all contract terms. Personnel representing all targeted areas of the plant should help devise terms that meet their area's needs. However, the supplier must also have an active role in defining contract terms as well. Before bids are requested, suppliers should be given the opportunity to review baseline data and study the facility. Chemical use in a plant is complex, covering many departments and processes. Therefore, chemical management contracts should not be developed without giving the supplier sufficient time to study and analyze chemical use. Mistakes can be costly for both sides. On one hand, suppliers are chemical experts who should be able ask the right questions and propose the appropriate range of services to define realistic contract terms. On the other hand, they must be given access to the necessary data and time to develop accurate estimates of chemical usage and costs.

GROWTH OF A SHARED
SAVINGS RELATIONSHIP

Shared Savings programs are guided by the plant's chemical management team. Here is a list of typical team activities.

- Coordinate the chemical management program through regular meetings (weekly, monthly or quarterly).

- Monitor and document chemical management program progress.
- Approve changes proposed by supplier personnel or plant personnel.
- Participate in the resolution of chemical-related problems.
- Investigate and resolve unexpected chemical incidents and liability problems.
- Recommend program improvements. (Mishra 1997a)

All the Shared Savings programs we studied evolved to address the unique circumstances, needs, and objectives of the respective chemical user and chemical supplier. The four case histories in this book offer contrasting pictures of Shared Savings relationships and their evolution from small single-chemical applications to plantwide comprehensive management programs. The most important point is that programs should grow and change through an evolutionary process rather than a planned expansion process.

Program evolution is driven by any of several factors. In some programs, the user expands the chemical footprint, potentially including all chemicals used in the plant. In other cases, the supplier's responsibilities tend to increase over time as the user's trust in the supplier's performance capabilities increases (see Figs. 22-2 and 22-3). When the financial terms of the contract are renegotiated programs tend expand to include additional services and profit-sharing strategies such as gainsharing. The most important factor contributing to program expansion is the trust that develops between both parties over time. The Shared Savings cycle is a self-perpetuating process that reinforces the aligned operational and financial interests of chemical user and chemical supplier.

After the initial year of a Shared Savings program, many companies request proposals from other suppliers and rebid the contract in order to assure that they are getting a fair price. It is interesting to note that most chemical users are skeptical of the lowest price bidders when these proposals are requested. They fear that the supplier is underbidding to get the contract. However, after the first rebidding of the contract, companies often choose to renegotiate with their current supplier rather than rebid future contracts. In fact, GM considers the rebidding of an existing contract a last resort, as both GM and the supplier have a vested interest in the relationship (Mishra 1997a). Renegotiation saves the supplier and the user considerable time and expense in writing and analyzing contract proposals. This valuable time can be devoted to improving the program.

IMPLICATIONS FOR THE MANAGER

Among companies using Shared Savings chemical management, most adopted the program in a time of crisis. They had reached a point where dramatic change was needed to reduce costs and improve performance. Resistance to change was overcome by the necessity to change. These companies now enjoy the benefits of their decision.

Unfortunately, in the absence of crisis, most companies do not recognize the need for improvement. Resistance to change is considerable and a sense of urgency is lacking. For this reason, most companies must prepare their road to Shared Savings. This means preparing plant personnel, establishing current baselines and the need for improvement, and identifying the supply base available to meet the Shared Savings needs of the plant. It may mean moving up the supply hierarchy in a series of steps, adding services, building trust, and changing the traditional chemical supply mindset. Most of all, developing a Shared Savings programs requires a champion—someone who promotes the program and sees it through the inevitable resistance that will be encountered.

CHAPTER

23

Shared Savings Contracts

*A good contract does not guarantee a good program,
but a poor contract will make a good program impossible.*

It is logical to assume that the key to a successful Shared Savings program is a well-written contract. However, we were told repeatedly by both the chemical users and chemical suppliers we interviewed that the key to success is in the working relationship established between the two companies and their personnel, not in the contractual clauses themselves. Few of the people we interviewed had even looked at the contract in the previous year, and some had never seen it. When we asked for copies of the contract, it usually took a while before suppliers as well as users could find a copy of it.

Everyone acknowledged that the contract established the ground rules for the relationship. The analogy of the contract to ground rules in sports is apt. Few baseball players ever read the rulebook that governs their game. A great game of baseball is a product of the talent and dedication of the players, not a great rulebook. Nevertheless, the rulebook establishes the basic expectations for both teams and the manner in which the teams interact or compete. Good rules do not guarantee a great game, but poor rules make a great game impossible. Similarly, a good contract is essential for both companies to achieve a profitable relationship. Many of the companies we studied learned this the hard way. Early Shared Savings contracts were poorly prepared. Though both parties had good intentions, problems with the contracts lead to confusion and conflict. In some cases, this resulted in failure of the Shared Savings program.

THE NATURE OF THE CONTRACT

We found that the structure and the content of contracts vary widely. Some are simple purchase orders with a standard set of predefined purchasing clauses that the supplier and user agree on and that address purchasing, delivery, management, liability, and other common characteristics of any supply relationship. Other contracts are large documents containing extensive baseline data, performance specifications, financial considerations, and so on. Still other contractual agreements take the form of multiple documents with a simple umbrella contract that refers to other, more detailed operating manuals and documents.

From these various Shared Savings contracts, we abstracted four components we believe to be essential components of any Shared Savings agreement:

1. Performance (including the chemical footprint, personnel, expectations, and fees)
2. Continuous improvement expectations
3. Miscellaneous provisions defining the relationship
4. Supporting documentation for performance assessment

We discuss each of these in more detail below. Regardless of the type of contract, however, in all cases the personnel involved stressed the importance of flexibility in the terms of the contract. Supplier and user personnel emphasized that despite their best efforts to include all critical specifications and expectations in the contract, circumstances and needs change. Giving consideration to these changes, the best way to fulfill the goals and objectives of a Shared Savings relationship is to adjust them to support the changes as well. It is important that the contract does not attempt to spell out in detail all the responsibilities and activities of the parties involved. The contract should be a guide that creates performance incentives and assurances that become the foundation of a Shared Savings relationship. Beyond that, it should be left to the personnel involved in the program daily to decide how best to meet the goals and objectives of the contract. In other words, the contract should give the relationship room to evolve and grow.

PERFORMANCE

The four key features of the contract that define performance are:

1. *Chemical footprint*—the chemicals and chemical systems covered by the contract
2. *Staffing and support*—the support personnel provided by the supplier
3. *Performance expectations*—user expectations regarding the chemicals and the chemical supplier
4. *Performance fees*—the financial consideration for the supplier

Chemical Footprint

The chemical footprint simply specifies the chemicals and chemical systems covered by the contract. Some companies prefer to focus on chemical systems because the actual chemicals used in those systems may be changed or eliminated as a part of the Shared Savings program. For example, it is preferable to write a contract covering "paint booth cleanup" as opposed to specifying "Solvent X" because the supplier may be able (or required in the contract) to improve the paint booth cleanup process, reduce volatile organic compounds (VOCs), or reduce costs by using another solvent or eliminating solvents altogether.

Staffing and Support

The level of staffing support provided by the supplier depends on the chemical footprint and performance expectations defined in the contract. Staffing is a key determinant in calculating the contract's performance fees. Staffing is typically specified in terms of the number of personnel to be provided, their qualifications or capabilities, and the hours per week they will devote to the chemical activities of in the plant. The chemical user usually provides office space, office equipment, or secretarial assistance to on-site supplier personnel.

Performance Expectations

Ideally, performance expectations should define the value that the chemical user expects from the chemical supplier. Value can be stated in terms of product value for the ultimate consumer, in terms of stockholder value, or in terms of value to the community, such as reducing the total life-cycle environmental impact of the company's products. However, few companies are currently able to measure such value, much less link it to the activities of their chemical supplier. Thus, most contracts address more tangible, in-plant measures of value. In time, though, performance expectations are likely to reflect higher measures of product value.

Performance expectations detail performance targets for the supplier. These targets commonly fall into five categories:

1. Product quality
2. Process performance
3. Services
4. Chemical characteristics
5. Miscellaneous

Well-defined performance expectations are not easy to develop. It is a time-consuming activity that requires the input of personnel throughout the plant. It can be tempting to list some desired services and the brand names of chemicals currently in use throughout the plant. However, poor performance expectation clauses produce poor performance. An understanding of the characteristics of good performance standards provides insight to their importance and the difficulty associated with defining them.

Product Quality

Product quality expectations are extremely important in contracts where the chemical footprint includes direct chemicals (those that become part of the product), such as paints, plastics, or coatings; or where the use of indirect chemicals can directly affect product quality, such as machining fluids. The most important performance characteristic of a chemical is its ability to produce a product that meets the quality standards of the chemical user. This performance standard is a primary measure of performance of the chemical supplier.

The best example of product quality expectations we encountered was at Chrysler's Belvidere assembly plant. Every car leaving the paint shop went through a Chrysler inspection. If the paint job did not pass the inspection, the paint supplier was not paid for that vehicle unless the problem could be traced by the Chemical Management Team to a Chrysler error. This performance standard made paint quality a primary performance responsibility for the paint supplier. The supplier's on-site personnel continuously monitored the painting process and worked to eliminate product factors that could lead to paint finish defects, even problems that were not related directly to the painting process, such as vehicle body preparation or equipment. Fewer paint problems meant that more first-quality cars were produced and the supplier earned more revenue. It also meant less rework expense, less waste, less disposal cost, etc., for Chrysler.

Other product quality expectations for chemicals might include characteristics such as rusting, spotting, cleanliness, or machining finish. Linking these expectations to supplier revenue aligns the performance and financial interests of the chemical supplier and the chemical user.

Process Performance

Process performance expectations are appropriate for both direct and indirect chemicals. An important measure of the performance for most chemicals is how problem-free a production process functions as a result of using a given chemical. Process performance expectations can be measured by many criteria. Listed below are a few examples:

- Maximum process downtime due to chemical malfunction
- Minimum tool life in a machining operation
- Maximum boiler and steam pipe corrosion rates
- Maximum VOC emissions
- Maximum waste byproducts
- Wastewater concentration limits

The value of many chemicals can be stated in terms of process performance and output. While these expectations are not always easy to establish, well-defined process performance expectations are essential to a successful Shared Savings relationship.

Services

Some chemical management services may be provided by a supplier as a component of product quality and process performance expectations. For example, monitoring and maintaining in-process chemical quality may be provided by some suppliers simply to assure that their product continually meets the product quality and process performance expectations of the contract.

However, most contracts involve an array of specifically requested services. Ultimately, the services included in performance expectations depend on the unique chemical needs of each plant. The services provided by the supplier are likely to expand and change over time. Many companies begin with limited services and increase the service level as the supply relationship matures and expands. In a number of cases we studied, the supplier sought these changes to achieve better control over the performance of their chemicals. In other instances, the user requested the expansion, because it found that the supplier was better at monitoring and controlling chemical performance than its own personnel were. Chapter 22 provides more details about the typical services included in a Shared Savings program and describes how these services evolved at General Motors and Navistar International.

Chemical Characteristics

Chemical characteristics are the functional aspects of a chemical—such as lubricity, viscosity, or pH—which define its performance. If product quality and process performance expectations are well defined, there may be no need for detailed chemical characteristics. For example, in a contract covering cutting fluids, expectations may be established for product finish quality, cutting tool life, rancidity, and other key product and process outcomes. Under these circumstances, it may be unnecessary or redundant to specify chemical characteristics. The supplier should have the flexibility to recommend other fluids as long as it can be demonstrated that product quality and process performance expectations will be maintained or exceeded by the new fluids.

However, many Shared Savings programs, particularly in their early years, may not have clearly defined product quality and process performance expectations to guide the selection and application of substitute chemicals. It may be necessary for the user to specify those chemical characteristics that are required to ensure process performance and product quality. Suppliers can and should be invited to participate in recommending and determining appropriate chemical characteristics.

It is important to avoid confusing the specification of chemical characteristics with brand specifications. When new equipment is purchased, the maintenance documentation often recommends specific brands of fluids to be used with the equipment. Equipment manufacturers include these brand-name recommendations because the fluids meet the necessary chemical characteristics required by the equipment. Then again, equipment operators or purchasing agents in the company may have favorite suppliers or chemical brands. While brand specifications may be acceptable at the start of a new Shared Savings program, they should not be considered appropriate substitutes for well-defined chemical characteristics. Chemical choices must ultimately be determined by product performance or process performance specifications. In the long run, brand specifications should be replaced by well-defined chemical characteristics. Eventually, chemical characteristics should be replaced with product quality and process performance expectations. These give the chemical supplier the flexibility to provide chemicals that produce excellent product quality without the constraints imposed by predefined chemical characteristics.

Miscellaneous

Shared Savings contracts may include a number of miscellaneous expectations, depending on the unique needs and circumstances of the companies involved. Some chemical users mandate chemical cost reductions annually. For example, a 5% reduction in the unit price paid for chemical management services might be mandated for each of the first three years of the Shared Savings contract. The mandated savings may take the form of a reduction in unit price *or* an equivalent amount of savings in other chemical-related areas. These other savings may include reduced scrap and rework, reduced wastewater treatment costs, reduced disposal fees, and reduced utility costs. The extent to which savings can be mandated as a performance expectation depends on the relative market power of the chemical user and chemical supplier and the extent to which immediate savings are possible through improved chemical management. For many companies, it is more realistic to strive for stable unit prices and rely on continuous improvement incentives (see next section) in order to produce additional savings.

Most contracts also specify reporting requirements for the chemical supplier. These requirements may include:

- The chemical user staff or committees to receive the reports
- Report contents or purpose
- Report frequency

In most Shared Savings programs there is a chemical management committee or similar group that oversees the contract. The chair of this committee is the lead contact person for the chemical user. This individual may be from Environment, Health, and Safety (EHS), Purchasing, Operations, or another division related to chemical use. The supplier provides written and oral progress regular reports to this committee. This interval is often monthly at the beginning of the contract but may become quarterly or even semiannual as the relationship matures. In addition, this committee often resolves unexpected contract performance problems.

The supplier may report to many other committees, depending on the scope of the chemical footprint. There may be a solvents committee, a paint shop committee, a quality committee, a safety committee, etc. In addition, there may be numerous plant personnel with whom supplier staff should communicate daily or at least weekly. These individuals might include EHS staff, production supervisors, lead line workers, paint shop manager, etc. Minutes from each of these meetings may be specified in the contract.

Necessary equipment related to chemical management should be specified in the contract. Often, the supplier provides equipment such as pumps, totes, filters, and computers. The contract must clarify who owns and maintains this equipment. The chemical user may also purchase equipment or other assets to be used by the supplier, including office space, telephone and fax capabilities, and so on. If the equipment is not specified in the contract, provisions should be included for equipment acquisition, payment, and operational responsibility.

Performance Fees

Performance fees specify how the supplier will be compensated for meeting performance expectations. These fees can drive both performance and continuous improvement. Most contracts use one or several of the following types of performance fees.

Fixed Fee

A fixed-fee arrangement is the simplest form of performance fee. It is commonly a fixed annual fee paid in monthly increments to the chemical supplier. The fee can be as straightforward as the historical annual chemical costs divided into twelve equal payments.

Unit Price

The most common performance fee is the unit price—a fee paid per unit of production. The most important reasons for using an unit price are that it directly compensates the supplier for changes in production volume and that it links chemical performance and supplier revenue.

Management and Service Fees

A management or service fee is paid to a supplier for special services that are not considered part of the core contract. This might include services such as computer-based chemical tracking, expanded health and safety education programs, and cost accounting programs.

Mixed Fees

Contracts may include a mix of the above fee strategies. Unit prices may be established to cover supplier costs that fluctuate with production levels and a fixed fee included to cover supplier equipment or chemical systems for which there are insufficient data to establish a link between costs and production levels. Finally, service fees may be layered on top of this for other chemical management services from the supplier.

The fee structure for each contract is driven by the unique needs of each plant and the capabilities of the supplier. However, the following guidelines can be used to help select the ideal fee structure. Given the importance of alternative fee structures, we devote a significant portion of the next chapter to explaining the most common approaches.

CONTINUOUS IMPROVEMENT

While the performance portion of the contract defines the basic expectations for performance and compensation in the relationship, the continuous improvement portion establishes incentives to drive performance *beyond* initial expectations. The emphasis of the contract is typically on cost reduction; this portion addresses value-adding opportunities for improvement, ranging from improved product quality to more visible corporate image. However, because cost savings are the easiest costs to measure, companies tend to focus on the financial incentives offered for continuous improvement in chemical management.

The financial incentive concept applied in the continuous improvement portion of the contract is simple: cost savings beyond those required under the performance expectations are shared by both parties. This motivates the supplier to generate performance savings beyond those required in the contract.

Two common mechanisms are used to create monetary incentives for continuous improvement. The first, *fee-based incentives*, is used to generate savings in the portion of the chemical cost iceberg covered by the performance expectations (see Fig. 20-1). The second, *gainsharing*, is used to drive savings in the remaining portion of the chemical cost iceberg as well as to stimulate the pursuit of savings opportunities in areas beyond the contract's chemical footprint. Below, we explain the use of fee-based incentives. Gainsharing is discussed in greater detail in the next chapter.

Recall from Fig. 20-1 that the chemical cost iceberg can be divided into costs that are covered by the performance expectations of the contract (the PE component) and the remaining costs, which continue to be the responsibility of the chemical user (the non–PE component). Fee-based incentives promote continuous reductions in the PE component of the iceberg.

The performance fees established in the performance section of the contract provide this incentive. Whether fixed fees, unit prices, service fees, or a combination of all three, performance fees are not directly tied to supplier costs. Thus, the supplier can increase profits by driving down the PE components costs.

In order to create an incentive for the chemical user to cooperate in driving down the PE component costs, some of these savings are shared with the chemical user. This sharing is typically accomplished via one of the following strategies:

- *Fee reduction*—Fees paid to the supplier can be reduced to reflect some or all of the cost savings realized by the supplier. A timetable and mechanism for fee reductions can be established that allow the supplier to retain all of the savings for a limited period of time and then pass some or all of the ongoing savings back to the chemical user through lower fees. Some companies adjust fees annually or on another regular time schedule. Other companies adjust fees when needed; the readjustment process is initiated when the supplier has driven down costs substantially.

- *Rebates*—A simple approach to sharing savings is for the supplier to pay or rebate fees to the chemical user for some portion of the savings realized. This can be done on a case-by-case basis or on a schedule. The proportion of savings to be rebated can be established in the contract or negotiated case by case as well.

- *Increased performance expectations*—An alternative approach to sharing savings is for the chemical user to increase the supplier's performance expectations. This usually involves additional chemical management services or chemical management responsibility for an expanded chemical footprint. Though fees remain fixed, the chemical user benefits from the increased supplier performance expectations.

In reality, most plants use some combination of these mechanisms. For example, fees may be periodically adjusted to reflect substantial and permanent reductions in chemical use. Rebates might be used to share savings in the short term, until fees are renegotiated. The supplier might assume additional responsibilities to compensate for miscellaneous smaller savings that have not been shared through rebates or fee adjustments. The plants we studied tended to focus on sharing savings of projects or programs that have a significant financial impact. Savings realized from small programs or projects were either allowed to go to the supplier or assumed to be compensation for small additional program costs periodically experienced by the supplier.

The important point is: *the greater the supplier's share, the greater the supplier's incentive.* This is probably one of the most difficult concepts of Shaved Savings for the chemical user to understand and accept. Even though the chemical user may have the market leverage to negotiate a greater share of the savings achieved through supplier ideas, it may not be in the chemical user's best interest to do so. While the minimum performance expectations are specified in the contract, the continuous improvement incentives create the significant cost-saving ideas in chemical management. The greater the share of savings given to the supplier, the greater the incentive to identify significant cost-saving ideas. Chemical users must overcome their natural tendency to maximize short-term earnings and focus on reinvesting more of those savings in their chemical supplier. The chemical user will benefit from the savings in the long term.

MISCELLANEOUS PROVISIONS

Liability

Many contracts contain the standard liability clauses that the chemical user would expect in any chemical supply contract. Additional language might include liability provisions unique to a Shared Savings relationship. For example, one plant specified that chemical leaks identified before the chemicals are unloaded from the supplier's trucks are assumed to be the responsibility of the supplier but that chemical leaks identified after delivery are assumed to be the responsibility of the plant, unless proven otherwise.

Companies reported that developing a procedure to determine the cause of chemical incidents that create liability was more important than attempts to detail the responsibility for the liability. The focus of the contract was to resolve and eliminate potential chemical liability, not to affix blame or responsibility.

Resolving Problems

Contracts should include provisions for resolving unanticipated events that have significant financial consequences for either user or supplier. Typical circumstances for which procedures should be prepared include:

- Unanticipated chemical use (due to equipment problems, worker error, spills, etc.)
- Errors in contract terms (usually due to inaccurate or incomplete information)
- Unanticipated price changes in the chemicals purchased from Tier 2 suppliers
- Major variations (up or down) in production levels (particularly plant shutdowns)

The contract should include a process for resolving disputes. There is no way to anticipate every problem that might arise, nor would such contractual clauses be practical. Instead, the contract should detail a process that both parties will follow to resolve problems fairly. For example, Ford uses an eight-step problem-solving process to gather information, identify the causes of a problem, arrive at a solution, and share the costs. The process is similar to those used by many companies that utilize teams to identify and resolve production problems.

Duration and Renewal of Contract

The duration of Shared Savings chemical management contracts varies from one year to three or even five years. The process for rebidding or renegotiating the contract at the end term is usually specified. Sometimes, the first renewal of a contract requires rebidding; however, after a second contract term is awarded, the contract is often *evergreen*—it is renewed with the supplier through a renegotiation process without requiring extensive rebidding documentation. This saves time and money for both the supplier and the user. It also strengthens the customer-supplier relationship that is critical for the long-term success of the Shared Savings program.

However, certain terms in the contract, particularly supplier fees, may be open for negotiation on an as-needed basis. This pricing flexibility allows both the chemical user and supplier to benefit from positive market changes and provides protection from adverse market changes. Methods and reasons for adjusting the financial terms of a contract should be specified in the contract.

Termination

Contracts typically include a divorce clause for termination of the agreement. Usually this involves a thirty-day written notice that does not require a reason or a cause. Provisions for the disposition of the chemical inventory at the time of termination are often included in the contract. The divorce clause is usually short and simple. As one chemical user stated it, "We did not enter into this relationship to get divorced. If we get divorced, we both lose."

Ramp-Up schedule

In the initial Shared Savings contract, responsibilities and activities may be phased in by the supplier over time. This schedule should be outlined in the contract to avoid misunderstanding, but again, interviewees emphasized the importance of flexibility. Frequently, both the supplier and the user did not understand the true complexity of a process until they attempted to improve it.

SUPPORTING DOCUMENTATION

Baselines

The baseline chemical usage rates, costs, and other key production data that are used to measure future performance are critical to the success of a Shared Savings relationship. Sometimes known collectively as the *costbook*, baselines represent everything that is known about a process and the chemical costs it generates at the initiation of the Shared Savings contract.

Baselines are critical because they are used by both parties to negotiate realistic performance expectations and performance fees. Though generating good baseline data can be time-consuming and expensive, it is an essential component of the contract for both chemical user and chemical supplier. Several Shared Savings programs have run into severe problems because of poor baseline data.

For each process covered in the contract, baselines typically include:

- Historical chemical use and cost
- Historical production levels for that process
- The extent to which chemical use and production levels are linked
- Other key costs or problems related to chemical use (waste generation, personal protective equipment, dermatitis, etc.)

Because of the importance of baseline data, both parties must be comfortable with its accuracy. In some cases, baselines are documented by the chemical user prior to beginning Shared Savings negotiations. In these cases, suppliers are offered the opportunity to verify baseline data before finalizing terms. In fact, one chemical user considered the chemical supplier's verification process when selecting a vendor. It rejected a vendor who did not critically analyze the baseline data given during a precontract tour of the plant. In other cases, suppliers develop baselines as part of the bid preparation process. In yet other cases, a preliminary contract was established and baselines were developed through a joint user and supplier effort during the first year of the program.

Operating Procedures

For some contracts, a manual of operating procedures, similar to those required for ISO 9000 certification, may be applied. This manual can specify important procedures to be followed, quality control measures, personnel responsibilities, and so on. The use of an operations manual assures that many important details are clarified and documented. An additional benefit of incorporating an external procedures manual through a reference in the contract is that it allows procedures to be improved and updated without requiring a formal revision of the contract.

IMPLICATIONS FOR THE MANAGER

The Shared Savings contract provides the rulebook by which the supply relationship operates. It is critical to the initiation of the program but should not be an important part of the day-to-day activities of plant and supplier personnel. That is, once the rules are established, there should be little need to refer to the rulebook in order to perform routine activities.

Nevertheless, a good contract is an important component of a successful program. It provides the means by which the chemical user clearly articulates its goals and what it values in chemical performance and supplier services. It also establishes the incentives for both sides, aligning their financial interests. Finally, it provides the means by which success or failure can be judged.

CHAPTER

24

Pricing

*The initial contract buys performance for the user today,
but the supplier's profit buys improved performance tomorrow.*

Price determines the level of supplier profit. Establishing a price in a Shared Savings chemical management program is different than product pricing in traditional chemical supply contracts. In traditional contracts, supplier profit is a necessary evil that the user seeks to minimize. However, in Shared Savings, supplier profit serves two purposes. First, it creates the incentives that drive supplier behavior. The financial incentives align the interests of chemical supplier and chemical user and result in activities that reduce chemical usage, reduce the chemical cost iceberg, and increase chemical performance. Second, supplier profit provides the capital the supplier needs for improvement, which, in turn, produces greater value for the chemical user in the future. That's a lot to expect from a pricing strategy!

In this chapter, we discuss the four common pricing mechanisms utilized in Shared Savings programs: unit price, fixed fee, management fee, and gainsharing. Of course, the proper level of supplier profit is a key consideration inherent in any pricing strategy. Before exploring specific pricing methods, it is necessary to understand the role of supplier profit in creating continuous chemical management improvements.

INVESTING IN SUPPLIER PROFIT

Supplier profit is not just an expense—it's an investment. The question is: how do you maximize your return on that investment and how much should you invest?

Chemical supplier profit is an investment in the future of the chemical user. Investing wisely can yield significant returns. The transition from seeing supplier profit as an expense to seeing supplier profit as an investment goes hand in hand with the transition from a traditional supply strategy to a Shared Savings strategy. When one is seeking chemicals at the lowest price, supply relationships are short term. A supplier may invest its profits in improvements that increase chemical quality and lower chemical costs, but there is little likelihood that the chemical user who pays those profits will receive the benefits. The time horizon of the relationship is simply too short.

In a Shared Savings relationship, however, the chemical user is buying chemical expertise, not chemicals. The time horizon of the relationship is measured in years or decades. Greater supplier expertise can lead directly to greater value for the chemical user. Only in a Shared Savings relationship can the chemical user expect to see a significant return from supplier profits.

Finding the Right Level of Supplier Profit

With any investment, maximizing return means investing the right amount in the right places. The same is true for supplier profit in Shared Savings. Too little profit can both diminish a supplier's incentives and starve it for capital. The supplier may be forced to find hidden ways to increase profits. Companies with experience in chemical management learned this lesson early in the first programs they implemented. The major North American auto companies now reject bids that they consider too low and unprofitable for the supplier. They learned from experience that underbid contracts result in poor supplier performance and hidden costs passed on to their plants.

On the other hand, too much supplier profit may reduce a supplier's incentives as well. If the benefits of future improvements become small relative to current profit levels, the supplier may have little incentive to invest in such improvements.

Though there is no formula for finding the right level of supplier profit, it is clear that the best suppliers are also the most profitable. Though paying a poor-quality supplier a high level of profit does not necessarily create a better-quality supplier, the opposite *is* true. If a top supplier earns only average profits, it will, in time, become an average supplier. Top suppliers in Shared Savings programs must develop and maintain the most expertise, the best resources, and the most dynamic research and development (R&D) programs. These qualities require the wise investment of profit.

Note, however, that higher supplier profits do not equate with higher program costs. A top supplier constantly seeks ways to reduce its own costs as well as those of the chemical user. A smart chemical user looks for ways to work with the supplier to reduce total chemical costs. By working together to eliminate costs from the supply chain, the ultimate customer receives greater product value. This increases product sales and financially benefits both the chemical user and chemical supplier. In other words, a smart chemical user helps the supplier increase profits by *lowering costs* rather than simply letting the supplier raise its prices.

Avoiding the I-Want-it-All Mindset

"I want it all" is typical of transactional supply programs, where price is viewed as a zero sum game—"The more money my supplier gets, the less money I get, and vice versa." This attitude can destroy a Shared Savings program.

The expression *I want it all* comes from an electroplating facility we studied. One of the facility's chemical suppliers, attempting to expand its relationship with the electroplater, developed a technology that would reduce chemical use, reduce waste, and produce a substantial cost savings for the electroplater almost immediately. The supplier would retain ownership of the technology and would install and maintain it in exchange for a portion of the savings generated by the new technology. However, the owner of the electroplating facility recognized that the supplier would increase its profits under the proposed arrangement. He responded, "No, I want it all," and insisted that the supplier sell the technology to him and hold no further stake in it. Even though he would have profited by the supplier's proposed program, he wanted the supplier's profits as well. After a prolonged and sometimes bitter negotiation, the supplier sold the technology to the electroplater. Later, when we spoke with the supplier representative, he said, "We'll never bring another waste minimization idea here again."

It was a shortsighted response by the owner. Though financially better off in the short term, he had cut himself off from the innovative technology and resources of the chemical supplier. The owner's transactional relationship with his supplier and his desire to grab all the short-term profits obstructed his ability to see the long-term profit opportunity the supplier was offering him. The owner used his market leverage to capture all the short-term profits for himself at the expense of long-term profitability.

Putting Supplier Profit in Perspective

To put supplier profit in proper perspective, it should be viewed in terms of the total cost of chemical ownership (the chemical iceberg) as well as the performance leverage of the chemical. These two factors determine the value of the supply relationship. Future improvements in these two factors are financed through supplier profit.

Consider, for example, a plant with an initial total chemical cost (iceberg) of $1,000,000. Part of this cost includes a Shared Savings contract with a supplier for $300,000, of which 10% ($30,000) is supplier profit ($270,000 is supplier cost). Now, assume that the supplier is able to cut its own costs by $30,000 to $240,000. This could occur by improving the efficiency of its services, improving the plant's efficiency of chemical use, or other means. (While such improvements typically produce savings for the plant as well, we assume, for this example, that only the supplier's costs are reduced.)

Now consider three scenarios for the reaction of the plant:

Scenario 1: Retain 10% supplier profit margin.
  If the plant adjusts the contract fee to keep supplier profit at 10%, then a new contract fee of $267,000 is established. This reduces the plant's total chemical costs to $967,000. Supplier profit declines from $30,000 to $27,000. While this is clearly advantageous for the plant (in the short term), it penalizes the supplier for its improvements. Such a mistake by the supplier is not likely to be repeated.

Scenario 2: Retain contract fee.
  If the plant retains the original contract fee, costs remain the same for the plant. However, supplier profit increases to $60,000, or 20% of the fee. This creates a strong incentive for the supplier to find further cost savings opportunities.

Scenario 3: Share the savings.

A third option is splitting the cost savings. If we assume that the savings are shared equally, the plant's costs decline to $985,000. Supplier profits increase to $45,000. The savings can be shared by changing the contract fees or through supplier rebates to the plant.

The implications of the first option are clear: if the supplier implements efficiency improvements, the reward is lower profits. A supplier will make this mistake only once. If, on the other hand, the supplier is able to keep some or all of the savings, it has an incentive to make further improvements.

Now imagine a scenario in which the supplier is able to reduce the plant's costs by $30,000, but not reduce its own. For example, the supplier might have found an alternative solvent with the same price and performance characteristics as the current solvent, but without special disposal requirements. The plant experiences reductions in disposal costs, but there are no financial benefits for the supplier.

Scenario 4: Plant retains all savings.

In this option, the plant retains all of the savings, reducing total chemical costs to $970,000, while the supplier's profits remain the same.

Scenario 5: 100% gainshare.

In this option, the savings are passed along to the supplier through a gainsharing provision in the contract. Supplier profits increase to 20%.

Scenario 6: 50% gainshare.

In this option, the savings are split evenly, reducing plant costs to $985,000 and increasing supplier profits to 15%.

Scenario 4 offers no financial incentive to the supplier to generate additional cost-saving innovations. Scenarios 5 and 6 do produce a financial incentive. In reality, gainsharing programs usually provide the supplier a portion of the savings for a period of time, after which the savings revert entirely to the plant. The supplier receives the short-term financial benefits while the plant receives the long-term financial benefits.

Supplier profit drives innovation, but this profit should be viewed in relation to the overall chemical cost iceberg. Compared to the baseline of 10% supplier profit, Scenarios 3 and 6 increase supplier profit to $45,000. This increase in profit represents only 1.5% of total chemical costs for the plant and is offset by a cost reduction of equal size. Within a year or two, the additional supplier profit reverts back to the plant. This is an attractive investment. Unfortunately, the I-want-it-all mindset can block such investments. Allowing the supplier to receive additional profits that "could" have gone to the plant is simply too much for some managers to tolerate. By taking all the additional profits, they cut themselves off from future improvements.

LINKING PRICES TO PERFORMANCE

One of the goals of Shared Savings chemical management is to pay the supplier for value-added performance, not just for chemicals. In situations where chemical performance can be clearly associated with product quality, unit prices offer a good mechanism for linking pay to performance. If the chemical does not perform satisfactorily and product quality does not meet quality standards, the supplier is not paid. Unit prices can also be useful in linking supplier pay to other aspects of chemical performance, such as equipment downtime. If equipment is down, production is reduced, thereby reducing supplier revenue. When supplier payment it linked to products through unit prices, the supplier has a financial stake in the performance of the chemical.

However, in many cases it is difficult to clearly link chemical performance with supplier fees. For example, it is difficult to link payment with performance outcomes such as tool life, pipe corrosion, or the cleanliness of paint booths, even with the use of unit prices. When using fixed fees or management fees, the link to performance becomes even more difficult because the fee is not directly linked to successful production of a quality product.

Most plants address this problem by establishing performance targets and monitoring performance outcomes closely. The supplier is expected to correct chemical performance problems immediately. In fact, the supplier should monitor chemical performance routinely and correct chemical-related problems before the intervention of the chemical user is required. Thus, while in some cases a unit price may be sufficient, most contracts include additional performance targets with a process for monitoring performance and correcting problems.

The fee structure for each contract is driven by the unique needs of each plant and the capabilities of the supplier. However, the following guidelines can be used to help select the ideal fee structure:

- Maximize the alignment of financial interests of chemical supplier and chemical user, with an emphasis on product quality and process performance.
- Assure fair financial benefits for both parties. Avoid conditions under which one party is forced to assume responsibility for losses due to factors unrelated to either party's control.
- Avoid fees that cannot be supported with data. Use fixed fees until sufficient data are available to establish reliable unit prices.

The key to developing a successful fee structure is for both parties to share the financial benefits of the chemical management program. Our research findings suggest that a significant amount of money can be saved through improved chemical management. By implementing a fee structure where both the chemical user and supplier can financially benefit by working together, the potential to realize these savings is greatly increased.

UNIT PRICES

The unit price fee structure is one of the more common payment strategies applied in Shared Savings chemical management programs. Unit prices offer several advantages. First, they are the easiest to link to performance, as illustrated above. Second, they are inherently linked to production, so there is no need to alter contract terms as production and chemical consumption change over time. For example, a significant increase in production levels increases supplier costs as more chemicals are consumed. Under a fixed fee arrangement, a new fee must be negotiated to compensate for the increase in production. However, under a unit price strategy, supplier revenue increases in proportion to production increases. A third advantage is that unit pricing has an important psychological effect on the supplier by constantly reinforcing its purpose at the plant—to help produce a quality product.

Selecting an Appropriate Unit

A unit price is simply a dollar amount paid to the supplier for each unit of product produced. It is common in the auto industry for the supplier to be paid for each vehicle manufactured. This strategy makes sense in applications that use chemicals in direct proportion to the number of units of product produced. In the auto industry, this includes chemicals such as paint, surface preparation chemicals, machining fluids, sealants, and solvents.

However, some chemicals may be only indirectly related to production rates, or may not be related at all. For example, water treatment chemicals used in the boiler house, wastewater treatment plant, or cooling towers may vary in use for many reasons other than fluctuations in production. For such chemicals, there are three possible options:

1. Continue to link fees to production of product, assuming that variances, on the whole, will average out.
2. Select an alternative fee structure (fixed fee or management fee).
3. Link fees to an intermediate production step rather than the final product.

The third option opens a wide array of alternatives. If a chemical is used in conditioning boiler water, the supplier could be paid per million pounds of steam produced rather than per vehicle produced. Chemicals used in paint detackification can be paid per hundred gallons of paint sprayed. Paints used in plant maintenance could be paid per square foot of area applied.

Collecting Appropriate Data

Establishing unit prices requires data. Most companies use from one to three years of historical data to determine prices. For the most basic analysis, the only data required are previous chemical expenses and previous production levels. For example, if a company spent $1 million dollars last year on a given chemical and produced 100,000 units, the baseline unit price would be $10/unit. For chemicals not related to production, a similar analysis can be performed. If a company spent $200,000 on boiler water chemicals last year and produced 100 million pounds of steam, the baseline unit price would be $2,000/million pounds of steam. (The actual contract unit price may be different from the baseline if the supplier provides price concessions or if additional revenue is provided to compensate the supplier for additional services.)

However, establishing an accurate unit price can often be more complex. Sometimes chemical usage is not well understood. One company we studied, for example, was trying to establish a unit price for paint detacki-fication and paint booth treatment chemicals by linking them to gallons of paint sprayed in the paint shop. However, the system seemed to be sensitive to a number of factors other than paint volume. It was difficult to produce a stable ratio of chemical use to paint use. In another example, Chrysler was attempting to establish unit prices on paint for its vehicles. However, the number of color and paint style combinations was quite large, and each combination required different volumes of paint. To further complicate the analysis, the different paint colors had significantly different costs. It required three years of study by both Chrysler and the paint supplier to develop reliable unit prices for painting. Eventually, specific unit prices were established for each paint combination.

When faced with complex chemical applications, it is tempting to abandon the effort and return to a more traditional pricing structure. The time, effort, and expense required to develop unit prices may appear too costly. However, plant managers must ask themselves whether they can really afford to continue their current level of ignorance about the process. In many cases, chemical systems are difficult to price because they are difficult to control. Without understanding the factors that influence system performance, the plant cannot hope to control them.

The companies we studied often involved their supplier in studying the systems and collecting data sufficient to establish unit prices. In some cases, supplier data collection was integrated into the initial contract. The contract can be initiated with an alternative fee structure, such as a fixed fee, based on the previous year's chemical costs. The supplier then uses the first year of the contract to collect needed data, analyze the data, establish baselines, and set unit prices.

FIXED FEES

While unit prices are linked to a unit of product, fixed fees are fixed per unit of time. Typically, they take the form of a flat monthly payment. An advantage of the monthly fixed fee is that it is simple, yet it offers a financial incentive for the supplier to reduce chemical usage and management costs.

Fixed fees can be particularly effective when program costs are not closely related to production, or when current data are insufficient to determine the association of chemical costs to production. For example, the cost of an oil recycling machine provided by the supplier is relatively fixed, regardless of production levels. Even if production is stopped, the machine continues to be an expense for the supplier that should be reimbursed by the user. The relationship between chemical use and production levels may not be well understood. It makes sense to use a fixed fee strategy until sufficient data are collected to determine an accurate unit price. Of course, if production levels have minimal variability, there may be little need to change to a unit price strategy from a fixed fee.

Fixed fees work in applications where chemical consumption does not vary significantly from month to month. In fact, fixed fees can be a superior alternative to unit prices where the actual unit product volume can change dramatically, yet overall chemical consumption is relatively stable. This can be the case, for example, in a machine shop where the type of part being machined (the product) changes frequently but the hours of machining and fluid use do not. Because significant, long-term changes in production levels can change chemical use and have financial implications for the supplier, fixed fees usually include trigger values or threshold levels for readjusting the fees to match either increases or decreases in production. Production fluctuations of 10–15% are typical trigger values.

Fixed fees are often easy to establish. Many companies simply begin with last year's chemical costs divided into twelve monthly payments. As with unit prices, the actual contract price may be different due to price concessions from chemical supplier or a premium paid by the chemical user for additional services, such as procurement or inventory management. The fixed fee may also cover the cost of special equipment provided by the supplier.

However, in some cases, appropriate fixed fees are difficult to establish. If a chemical system is not well understood or out of control, the previous year's chemical costs may not be a good predictor of future chemical costs. Similarly, changes in the system, such as the introduction of a new product or a new process technology, can make future chemical use unpredictable. The contract must be written to ensure that both parties share and provide a mechanism for resolving inappropriate fee structures.

MANAGEMENT FEES

Management fees are sometimes referred to as *service fees* or *pass-through* fees. Essentially, the cost of the supplier's chemicals is passed through to the chemical user and an additional fee is added to cover supplier services.

Many suppliers prefer management fees because they create another revenue opportunity while carrying limited risk. Supplier profit is not dependent on accurately estimating chemical use or successfully implementing efficiency improvements. Many chemical users are also comfortable with management fees. They are more consistent with the traditional sales approach of purchasing chemicals and allow the user to more easily assess the level of supplier profit, as it is not buried in the chemical price. It is simply a cost-plus system. The plant determines the cost of chemicals, adds the cost of services, and includes a level of supplier profit.

Unfortunately, management fees do not offer the performance incentives that are inherent in unit price and fixed fee strategies. The supplier also has little incentive to find ways to reduce chemical use, as this produces no added revenue. Thus, despite its initial attractiveness for many chemical users and chemical suppliers, management fees should not be used as a long-term strategy in Shared Savings programs.

Instead, such fees are best used on a short-term basis. As the footprint of the supplier expands into new areas, the use of management fees can limit supplier risk while the new chemical systems are being studied. Once sufficient data are collected to understand and control the systems, it is best to shift to a fixed fee or unit price.

However, a few companies are using a modified version of the management fee that can provide many of the benefits offered by unit prices and fixed fees. Sometimes called the "at-risk management fee," this approach links the supplier's fee to target cost reductions. For example, a supplier might agree to produce annual savings equal to 5% of the chemical supply contract. These savings can take the form of chemical volume reductions or other savings, such as utilities, waste disposal fees, and so on. If the annual savings fall short of the 5% target, the difference is made up through a reduction in the supplier's fees. In this way, the supplier has a financial incentive to reduce chemical volumes, albeit only to the level specified for annual savings.

GAINSHARING

Unlike the fee arrangements discussed up to this point, gainsharing is not an ongoing fee; it is a mechanism for sharing benefits on a case-by-case basis. With respect to the chemical cost iceberg (Fig. 20-1), gainsharing applies to chemical or other operating costs outside the performance expectation portion (the non–PE component). For example:

- The supplier developes a new paint formulation that cures at lower temperatures, reducing energy costs for the curing ovens.
- A supplier's efforts to improve machining fluid management improves finish quality and reduces scrap.
- A supplier identifies a superior coolant, increasing machining tool life.
- A supplier is able to reduce delivery and inventory volume, thereby reducing chemical spills and associated chemical disposal costs.

In each case, the actions of the chemical supplier create savings for the chemical user but not the chemical supplier. All of these savings lie in the non–PE component of the chemical cost iceberg, beyond the supplier's performance expectations. Gainsharing is a means of giving the supplier an ongoing incentive to find more such savings.

Gainsharing programs are not complex. The proportion of savings that is shared with the supplier can be established in advance or negotiated case by case. Many companies use a 50% distribution of savings for small amounts and reduce the supplier's percentage as the amount of savings increases. For ideas that produce savings over a period of time, defining a time limit is also necessary. The supplier's savings participation typically expires from one to five years after savings are realized.

Documenting savings can be difficult. Generally, the burden of documentation is on the supplier. The supplier must estimate the savings from the proposed project and identify how the actual savings can be measured. If approved, the idea is implemented and the savings are monitored according to the plan. If actual savings are substantially different from projected savings, the gainsharing formula may be modified.

In some cases, the financial interests of chemical supplier and chemical user may conflict. For example, the supplier may be aware of a new clear-coat paint that can be cured at lower temperatures, producing significant energy savings for the chemical user. However, if the paint has a higher cost, its use would reduce supplier profits because the supplier is paid on a unit price basis. In another scenario, the supplier may have a more expensive cleaning agent that is less toxic than the current cleaner and does not require expensive personal protective equipment for employees. In the absence of gainsharing, the supplier would be financially hurt by these innovations. *With* gainsharing, however, the supplier would be compensated for the benefits enjoyed by the chemical user, aligning the interests of both.

Of course, it is impossible to capture and share all benefits achieved through a Shared Savings mechanism. Many benefits are soft or intangible. For example, reductions in chemical toxicity can significantly improve working conditions for employees and increase productivity. This may be an enormous benefit for the chemical user, but such benefits are not easily quantified and shared.

Many of the chemical users we interviewed commented on the advantage of having new eyes in the plant to look at old problems from a different perspective. The supplier brings a background, expertise, and ideas to the plant that the chemical user does not possess. As discussed in chapter 12, this aspect of the Shared Savings relationship may be the single most significant source of savings offered through such programs in the long term.

IMPLICATIONS FOR THE MANAGER

Obtaining the benefits of a Shared Savings chemical management program requires a significant change in the mindset of the chemical user. In short, supplier profit must be viewed as an investment rather than an expense. Like any other investment, it must be managed for optimum long-term return. Too *much* profit, and the incentive to improve is diminished. Too *little* profit, and the supplier will be unable to sustain its level of innovation and expertise, and may even turn to hidden means of increasing its profits.

Fortunately, any reasonably priced program can begin to generate savings for both chemical user and chemical supplier. Over time, pricing strategies can be refined to optimize incentives. The greatest threat to the ongoing success of a Shared Savings program is the chemical user's temptation to want it all—cutting off the program's ability to continuously reduce costs and improve performance.

25

The Future of Shared Savings

We began this book by introducing a dreamworld in which a company's chemical supplier:

- Continuously searches for ways to *minimize* the amount of chemicals the company requires to produce a quality product or service
- Helps to improve company operations to *reduce* the amount of chemicals wasted in production
- Provides the company with additional environmental management *resources* to increase environmental *performance* and generate *cost savings* for plant management
- Collects data on chemical usage and monitors chemical inventory to support the company's environmental *reporting* responsibilities
- Demonstrates how the company's environment, health, and safety activities contribute to core business goals

By now, the reader should recognize that this dreamworld is a reality for those companies that have successfully implemented Shared Savings chemical supply programs. For these companies, the Shared Savings model of chemical supply has demonstrated its inherent superiority to traditional chemical supply models and that to return to the old way would be a step backward. Yet, despite its advantages, Shared Savings is not being rapidly adopted by other companies or other industries. For those companies that understand and successfully implement Shared Savings, the opportunities for competitive advantage are unlimited.

In this chapter, we attempt to explain why most companies have been slow to adopt Shared Savings and how early adoption can increase competitive advantage. The fact that Shared Savings is still in the early stages of diffusion in industry provides a excellent competitive advantage opportunity for those companies who aggressively pursue Shared Savings now. As Shared Savings moves up the diffusion curve with more and more organizations adopting the strategy, those companies that have Shared Savings programs in place will have a significant competitive lead over the companies that haven't. Given the time it takes to develop a Shared Savings mindset, the companies that implement Shared Savings earlier in the diffusion process will benefit from the strategy while other companies are still in the implementation phase.

HOW NEW IDEAS SPREAD

An innovative technology or idea typically diffuses through a population in a predictable pattern. Analysis of this pattern, known as *diffusion of innovations research*, began in the study of agricultural innovations but has proven to be extremely useful in marketing most products or ideas (Rogers 1995). Diffusion researchers documented that individuals approach opportunities for change (innovation) from different mindsets. Fig. 25-1 is an illustration of the classic *cumulative adoption curve* for diffusion of an innovation over time. Innovations have a relatively slow adoption rate in the early stages, followed by a phase of rapid adoption until the rate of adoption eventually levels off (for a detailed discussion of the adoption curve and its causes, see Rogers 1995). The *first-time adoption curve* identifies the characteristics of people or businesses that adopt an innovation in each stage of adoption curve. These characteristics are summarized in Table 25-1.

Our research suggests that the diffusion of Shared Savings is currently in the early adopter stage. It is gaining ground among mainstream companies seeking a competitive advantage, but it is not perceived as an industry standard or trend. The companies that adopt Shared Savings tend to be risk averse, but they operate in very competitive markets. These companies are open to change and innovation, providing it is profitable. These companies continually seek to create a competitive advantage, as they must

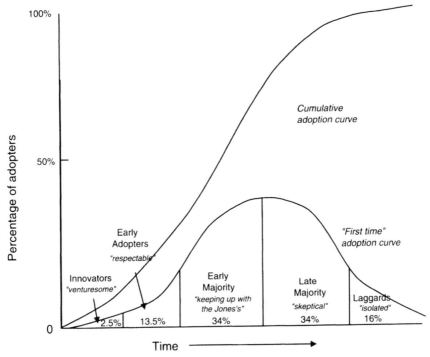

Figure 25-1. The pattern of innovation diffusion over time (Bierma and Water-straat 1995).

Table 25-1.
Characteristics of Adopters by Category
(adapted from Bierma and Waterstraat 1995; and Rogers 1995)

Innovators	• Characterized as "venturesome" and "risk takers" • Interested in innovation • Possess necessary resources
Early adopters	• Are close to mainstream • Are open to change and innovation • Characterized as risk averse
Early majority	• Are well informed • View new innovations as too risky • Fear being left behind • Follow major trends
Late majority	• Delay until innovation is necessary • Driven by social norms and standards
Laggards	• Are socially isolated • Focus on past practice • Possess no future vision

be leaders in both product quality and value just to hold market share. They are finding that Shared Savings provides them with additional ideas and resources to improve the quality and value of their products.

Diffusion of innovation theory proposes that the greatest competitive advantage is realized by those who adopt the innovation during the early adopter stage of the diffusion curve. Ultimately, most businesses adopt an innovation to avoid being left behind once the rest of the industry has captured the benefits of the innovation. The greatest benefits gained from the competitive advantage produced by Shared Savings will be experienced by companies that move now.

THE GREATEST BARRIER TO DIFFUSION: A NEW WAY OF THINKING

Through our interviews, we determined that the single greatest barrier to Shared Savings is the traditional supplier relationship mindset. This mindset is held by both chemical suppliers and chemical users. It is the basis for suppliers' chemical sales strategies as well as chemical users' purchasing strategies. Because Shared Savings requires that the traditional mindset be abandoned, it can appear too risky for many companies.

Yet the need for most companies to implement and benefit from a competitive advantage demands that the elements of the traditional supply mindset be challenged. This challenge has led a number of firms to adopt what we call the *Shared Savings mindset* and thereby benefit from a significant increase in performance and product value. In Table 25-2, we compare the common elements of these two supply strategy mindsets.

The Shared Savings mindset has helped companies to:

- Minimize total cost for value chain companies.
- Maximize total quality for value chain companies.
- Maximize efficiency of the value chain companies.
- Maximize new market opportunities to the value chain companies.
- Increase the competitive advantage of the value chain companies.

It is evident that abandoning the traditional supply mindset is the most difficult step that companies must make to implement superior supply relationships. Yet the competitive advantage offered by Shared Savings will

Table 25-2.
Comparison of Traditional and Shared Savings Supply Mindsets

Traditional Supply Mindset	Shared Savings Mindset
Supplier profits from increasing chemical volume.	Supplier profits from increasing customer value.
Chemical user seeks to minimize purchase price.	Both supplier and use seek to minimize total cost.
Chemical user benefits by shifting costs to others in the value chain.	Both the supplier and user benefit by eliminating costs from the value chain.
Creates lack of trust in value chain members.	Builds trust with other members of the value chain through shared incentives.
Supplier-customer relationships are defined as a zero-sum game.	Supplier-customer relationships are defined as an opportunity with unlimited capacity to create value and profits.
Supplier and user focus on short-term profits.	Supplier and user focus on long-term profits.

benefit only those companies that are able to achieve the mindset transition before their competitors do.

BARRIERS TO WIDER DIFFUSION

While the traditional supply mindset is the primary barrier to the adoption of Shared Savings, technical barriers currently limit the diffusion into certain industry sectors. Overcoming these barriers requires innovative approaches to Shared Savings.

Chemical Account Size

One of the major barriers is the size of the chemical account. Chemical accounts in excess of several million dollars are typical of today's successful Shared Savings programs. These accounts can support on-site supplier personnel who actively manage the account as well as pursue additional opportunities to reduce chemical use and costs in the plant. The lower limit for chemical account size depends on the inherent profitability and efficiency of the account itself. Accounts composed mainly of small-margin chemicals, such as commodity chemicals, would probably require an account value in excess of several million dollars to support one on-site supplier representative. On the other hand, an account using high-margin proprietary chemicals might support a Shared Savings program of one million dollars or less. How efficiently these chemicals are used also affects the potential value of the account. A factor closely related to overall value and efficiency of the account is number of individual chemicals used within the account. A large number of small-volume chemicals can make it difficult to support a Shared Savings program, but some suppliers have successfully consolidated chemical inventories that were out of control.

Reducing this barrier requires innovative use of personnel and technology. Advances in telecommunications and computing technology allow suppliers to remotely monitor and manage many aspects of small accounts. Other advances in equipment also allow Shared Savings programs to work on a more limited scale. For example, we are aware of one major supplier of aqueous cleaners who formed a partnership with a manufacturer of membrane filter systems. Working together, they designed the chemical cleaners and filters to work effectively at a plant site, cleaner quality was improved, and a reduction in cleaner waste was achieved. As a result, these two companies are able to offer customers a unique Shared Savings program. Instead of buying cleaner, the plant pays a user fee for each gallon of cleaner that cycles through the parts washer. An alternatively payment strategy could pay the supplier a unit price for every part cleaned by the customer. In either case, it is in the supplier's interest to maximize quality and minimize waste. This arrangement is appropriate for accounts of $500,000, and probably less.

Variable Production Barriers

A second barrier to successful Shared Savings implementations is extreme variability in chemical use. This problem typically occurs in operations where there are large fluctuations in production levels or frequent changes in the type of product manufactured. Job shops usually present some of the biggest challenges for successful Shared Savings programs. The products and production levels constantly change, often in unpredictable ways. Creating successful Shared Savings programs under such circumstances requires creative approaches to staffing, applying technology, and determining contractual terms.

IMPROVING SHARED SAVINGS

Despite the success of the Shared Savings strategy, it generates new problems that must be addressed. Several of the most common problems we encountered during our research are discussed below.

Tier 2 Suppliers

"You don't ever want to be a Tier 2 supplier," remarked one supplier representative with both Tier 1 and Tier 2 experience. The problem, as he defined it, was that the Tier 2 suppliers literally work *for* the Tier 1 supplier. While this model was designed to have the Tier 1 supplier manage all chemical-related functions so the chemical user could focus on manufacturing instead of managing chemical suppliers, it has drawbacks. In most cases, the Tier 1 and Tier 2 suppliers are competitors. In some cases, they are *direct* competitors, offering similar products and services.

The Tier 1 supplier is expected to coordinate all chemical-related activities and function as a buffer between the chemical user and the Tier 2 supplier. When the Tier 2 supplier suggests new product ideas or other innovations, they are passed through the Tier 1 supplier to the chemical user. This creates a conflict of interest for the Tier 1 supplier. New ideas that reduce chemical costs increase the Tier 1 supplier's profits. However, new ideas presented by a Tier 2 supplier could build the competitive strength of the Tier 2 supplier that is competing in the same markets as the Tier 1 supplier. In time, the Tier 2 supplier might be able to replace the Tier 1 supplier at that plant. In fact, many of the Tier 2 suppliers we interviewed said that one of the reasons why they wanted a Tier 2 presence was to demonstrate to the chemical user that *they* should be the Tier 1 supplier. A related problem is that some Tier 1 suppliers do not extend the Shared Savings benefits to Tier 2 suppliers. The Tier 1 supplier has a transactional sales relationship with the Tier 2 supplier, buying on a dollars-per-pound basis. The Tier 1 supplier applies pressure on the Tier 2 supplier to reduce the cost of chemicals in order to justify its Tier 1 position to the chemical user.

Maintaining Proper Incentives

Financial incentives in Shared Savings contracts are used to align the incentives of chemical user and chemical supplier. Sometimes defining the incentives requires a bit of fine-tuning and creativity. In a few instances, suppliers reported that the fee structure eliminated the chemical user's incentive to work with the supplier to reduce chemical volumes. Because the chemical user no longer had to pay for the chemicals used under a unit pricing model, the supplier had to bear the cost of chemicals regardless of the amount of chemical used. The chemical users had no incentive to take steps to conserve or reduce chemical usage.

Suppliers indicate that this occurs even though chemical users *do* benefit from reduced chemical usage. They receive benefits not only through sharing the supplier's savings but also through reduced costs for waste disposal, compliance, training, liability, and so on. However, as one supplier put it, "Operating managers are used to driving down chemical costs, not chemical volumes." This narrow focus on the tip of the iceberg can be difficult to change.

In cases where a fixed fee or unit price is insufficient to motivate managers to reduce chemical volumes, using a dollars-per-pound fee together with unit pricing or a fixed fee may produce better results. This method ensures that the chemical user experiences the cost impact of poor chemical management. One plant used another innovative approach. It negotiated significant up-front cost savings with the supplier but allocated these savings to various departments only as they made progress in reducing chemical volumes.

Changing Tier I Suppliers

Shared Savings relationships continually expand and evolve in most successful applications. The chemical footprint usually changes dramatically over time. Sometimes the footprint can change so significantly in terms of chemicals and volume that the chemical user decides to change Tier 1 suppliers to one that better meets the plant's chemical needs. In the auto industry, this is beginning to happen in some plants that began their Shared Savings programs with indirect chemicals, such as water treatment and paint detackification, but are expanding to include direct chemicals, such as paints. Paints represent a far greater percentage of the total chemical budget than do water treatment and paint detackification chemicals for most automobile manufacturers, and they have far greater impact on product quality. It makes sense for plants to switch to a Tier 1 supplier whose core business is paints. However, this can be a difficult transition. It shifts the current Tier 1 supplier to a Tier 2 position, which is not desirable for most chemical suppliers, as previously discussed. The current Tier 1 supplier probably made significant investments in personnel and equipment in the plant and developed valuable relationships and operational insights that may be lost as a Tier 2 supplier. It remains to be seen how these transitions play out in the long term for both the chemical user and supplier.

IMPLICATIONS FOR THE MANAGER

The move to Shared Savings is not an easy one. Most importantly, it requires that a company let go of the traditional supply mindset. In many companies, the traditional approach to chemical supply shaped the very structure of the organization, from Purchasing to Waste Management. It also influenced the budgets, relationships, and even status of departments and personnel throughout the company. Changing the approach to chemical supply can appear threatening to many people.

You can either be part of the resistance or a catalyst for change. Every plant we visited stressed the need for a champion to promote Shared Savings and guide its implementation. You can be among the first to question the traditional supply mindset and to explore the Shared Savings mindset, and can even become the champion of a Shared Savings chemical supply strategy. The opportunities are limitless. It is time to stop buying chemicals and start buying chemical performance. Value, not volume, produces competitive advantage.

GLOSSARY

Adaptability: The ability to adapt quickly to changing conditions. Adaptability is one component of a company's *capability*.

Ancillary operations: Operations that produce chemical and nonchemical inputs for other operations, or that manage waste outputs from other operations.

Anticipation: The ability to anticipate changes in customer needs and market conditions. Anticipation is one component of a company's *capability*.

Application costs: Costs directly associated with the use of a chemical in an operation.

Application services: Supplier services associated directly with the use of a chemical in an operation.

Baseline: Documentation of the chemical use and costs at the beginning of a contract (sometimes called the *costbook*).

Business value: The value created by a business for its stakeholders, which, at a minimum, include its owners and customers.

Capability: A company's ability to generate value in the future.

Chemical Beast: The total negative impact of chemical use in a company on the creation of business value.

Chemical Beast growth cycle: The cycle of distrust, conflict, and threat—common to traditional supply programs—that leads to a growing Chemical Beast.

Chemical characteristics: Attributes of a chemical required to achieve a specified level of performance. Chemical characteristics are one example of *performance expectations*.

Chemical cost iceberg: The total cost of chemical usage for a company.

Chemical life cycle: The steps in chemical use, from ordering to disposal.

Chemical management: Chemical supply relationships characterized by a wide array of supplier services and incorporation of the supplier into chemical problem-solving activities.

Chemical management fee: A fee paid to a chemical supplier to perform specified chemical management activities (also called a *management fee* or *service fee*).

Chemical performance leverage: The increase in business value resulting from a specified improvement in the performance of a chemical.

Chemical/process web: The complex interaction between chemicals and processes in which they are used.

Consumer value paradigm: The perspective that business value is created only when value is realized by the ultimate consumer. This perspective suggests that it is the performance of chemicals, not simply their supply, that creates value, and that costs must be eliminated, not simply shifted between members of the *value chain*.

Costbook: See *Baseline*.

Cost of chemicals: For a supplier, this is the cost of chemicals supplied to an account.

Cost of marketing: For a supplier, this is the cost of obtaining new accounts. It also includes the cost of retaining existing accounts, other than chemical and service costs.

Cost of services: For a supplier, this is the nonchemical cost of meeting the performance expectations of the chemical user, including logistic, EHS/compliance, and application costs.

Cost shifting: Moving costs from one member of a *value chain* to another member.

Cross-functional integration: The coordination of activities across functional boundaries in a company.

Dollars per pound: A chemical supply contract fee whereby the supplier is paid for chemicals on a volume or weight basis (pound, gallon, drum, etc.).

EHS: Environment, health, and safety. Programs concerned with environmental management and the protection of workers.

Entrepreneurial energy: An increase in productivity and innovation that occurs when an individual's interests coincide with the core business of his or her employer.

Evergreen contract: A supply contract that is expected to be renewed as long as both parties are satisfied with performance.

Fee structure: The method employed by the chemical user to compensate the chemical supplier.

Fixed fee: A fee structure, common to Shared Savings chemical management programs, whereby the chemical supplier is paid a fixed amount per month or other time period.

Footprint: The range of chemicals covered by a chemical supply contract.

Gainsharing: A fee structure, common to Shared Savings programs, whereby a portion of the chemical user's benefits are shared with the chemical supplier.

Indirect chemicals: Chemicals that are used in operations but do not become part of the product.

Intelligence: The possession of useful knowledge about business operations. Intelligence is one component of a company's *capability*.

Learning and innovation: The ability of an organization to learn or invent new products, processes, technologies, and strategies. Learning and innovation is one component of a company's *capability*.

License to operate: Compliance with regulations applicable to a company.

Limited chemical management: A chemical management program in which the chemical user buys chemicals on a price-per-volume basis.

Logistic costs: Costs associated with acquiring and handling chemicals.

Logistic services: Supplier services associated with acquiring and handling chemicals.

Mixed fees: A combination of fixed fee, unit price, or gainsharing fee structures in a Shared Savings chemical management contract.

Non–PE component: That portion of the chemical cost iceberg not covered by the *performance expectations* of a Shared Savings chemical management contract.

PE component: That portion of the chemical cost iceberg covered by the *performance expectations* of a Shared Savings chemical management contract.

Performance expectation: The contributions to value that a chemical user expects from a chemical supplier in a Shared Savings chemical management contract.

Process performance expectations: A specified level of performance for an operation. Process performance expectations are one example of *performance expectations*.

Product quality expectations: A specified level of quality in product. Product quality expectations are one example of *performance expectations*.

Service expectations: A set of responsibilities or activities to be performed by the chemical supplier. Service expectations are one example of *performance expectations*.

Shared Savings chemical management: A chemical management program in which chemical supplier revenue is not linked to the volume of chemicals supplied and in which the supplier has a financial incentive to reduce the *total cost of ownership* for chemicals.

Shared savings cycle: The cycle of mutual interest, trust, and cooperation—common to Shared Savings supply programs—that leads to control of the Chemical Beast.

Tier 1 chemical supplier: A chemical supplier that contracts with the chemical user and that oversees the supply of chemicals from lower-tier suppliers.

Tier 2 chemical supplier: A chemical supplier that contracts to provide chemicals to a Tier 1 chemical supplier. A Tier 2 chemical supplier may oversee the supply of chemicals from lower-tier suppliers.

Total Cost of Ownership (TCO): The total cost associated with a chemical throughout its life cycle in the plant.

Transactional chemical supply relationships: Chemical supply relationships focused on the exchange of chemicals for money.

Unit price: A fee structure, common to Shared Savings programs, whereby the chemical supplier is paid a fixed amount per unit of product produced.

Value chain: The chain of companies responsible for adding value to a product, from the supply of raw materials to the delivery of finished product to the consumer.

Volume conflict: The conflict created between chemical supplier and chemical user by the purchase of chemicals on a price-per-volume basis. The supplier profits from increasing chemical volume while the user profits from decreasing chemical volume.

BIBLIOGRAPHY

Bernath, R., 1997. Chemical management best practices. Paper presented at Chemical Strategies Partnership Workshop, Chemical Strategies Partnership, in San Francisco, 20 November.

Bierma, T.J., and F.L. Waterstraat. 1995. Marketing pollution prevention. *Pollution Prevention Review* 5:27–40, Spring.

Bierma, T.J., and F.L. Waterstraat. 1996. P2 assistance from your chemical supplier. *Pollution Prevention Review* 6:13–23, Autumn.

Bierma, T.J., and F.L. Waterstraat. 1997a. *Innovative chemical supply contracts: A source of competitive advantage.* Champaign, Ill.: Illinois Waste Management and Research Center. Report # TR-31.

Bierma, T.J., and F.L. Waterstraat. 1997b. Pollution prevention assistance from chemical suppliers? The Chemical Management Service contract and the surface finishing industry. Paper presented at the Eighteenth Annual AESF/EPA Conference on Pollution Prevention and Control for the Surface Finishing Industry, American Electroplaters and Surface Finishers Society, in Orlando, Florida, 27–31 January.

Bierma, T.J., F.L. Waterstraat, and J. Ostrosky. 1998. Shared Savings and environmental management accounting: Innovative chemical supply strategies," In *The green bottom line: Environmental managerial accounting—Current practice and future trends,* edited by M. Bennett and P. James. Sheffield, UK: Greenleaf Publishing.

Broe, B. 1997. Implementation of chemical management at Nortel semiconductors. Paper presented at Chemical Strategies Partnership Workshop, Chemical Strategies Partnership, in San Francisco, 20 November.

Craig, M.R. 1998. Hazardous materials management implemented by Delta Air Lines. Paper presented at Chemical Strategies Partnership Workshop, Chemical Strategies Partnership, in San Francisco, 21 May.

Daniels, H. 1998. Chemical management: A total systems approach. Paper presented at Chemical Strategies Partnership Workshop, Chemical Strategies Partnership, in San Francisco, 21 May.

ENDS Report, 1997. Nortel: Shared savings for chemicals and waste reduction. Report #267. Environmental Data Services Ltd. (UK), April.

Grieco, P.L., and M. Pilachowski. 1995. *Activity-based costing: The key to world-class Performance.* Palm Beach Gardens, Fla.: PT Publications, Inc.

Heller, M., D. Shields, and B. Beloff. 1995. Environmental accounting case study:

Amoco Yorktown refinery. In *Green ledgers: Case studies in corporate environmental accounting,* edited by D. Ditz, J. Ranganathan, and R.D. Banks. Baltimore: World Resources Institute.

Imai, M. 1986. *Kaizen: The key to Japan's competitive success.* New York: McGraw-Hill.

Johnson, D.R., and P.W. Burnap. 1995. Majestic Metals, Inc. *Pollution Prevention Review* 5:65–74, Autumn.

Kapoor, V., and A. Gupta. 1997. "Aggressive sourcing: A free-market approach. *Sloan Management Review* 39:21–31.

Knoblock, M. 1998. "Chemicals management—Best Practices: General Motors Truck Group, Fort Wayne, Indiana. Paper presented at Chemical Strategies Partnership Workshop, Chemical Strategies Partnership, San Francisco, 21 May.

Krampe, R. 1998. Chemical management systems: Process and opportunities. Paper presented at Chemical Strategies Partnership Workshop, Chemical Strategies Partnership, San Francisco, 21 May.

Lagoe, D.J. 1996. Alcan Rolled Products Company. *Pollution Prevention Review* 6: 51–57, Winter.

Lewis, J.D. 1990. *Partnerships for profit: Structuring and managing strategic alliances.* New York: Free Press.

Loren, J. 1996. "Weyerhaeuser implements new chemical management program. *Total Quality Environmental Management* 5:101–106, Spring.

Machinery Market. 1992. Cool savings: Benefits for a motor vehicle manufacturer as the result of close partnership with a supplier of metal-working lubricants. 11 June.

MacNabb, R.R. 1997. *Outsourcing + Insourcing = Best Sourcing (ER-426).* Manufacturers Alliance, Arlington, VA.

Mishra, P.N. 1997a. Chemical management best practices. Paper presented at Chemical Strategies Partnership Workshops, Chemical Strategies Partnership, in San Francisco, 20 November.

Mishra, P.N. 1997b. Personal communication with author. 5 August.

Mishra, P.N. 1998. Chemicals management. Paper presented at Chemical Strategies Partnership Workshop, Chemical Strategies Partnership, in San Francisco, 21 May.

Moore, J.F. 1996. *The death of competition.* New York: HarperCollins.

Moshure, P. 1998. Chemical and gas management program. Paper presented at Chemical Strategies Partnership Workshop, Chemical Strategies Partnership, in San Francisco, 21 May.

Poirier, C.C., and W.F. Houser. 1993. *Business partnering for continuous improvement.* San Francisco: Berrett-Koehler.

Reid, B. 1997. Chemical Management at General Motors of Canada Ltd. Paper presented at Chemical Strategies Partnership Workshop, Chemical Strategies Partnership, San Francisco, 20 November.

Riggs, D.A., and S.L. Robbins. 1998. *The executive's guide to supply management strategies.* New York: American Management Association.

Rogers, E.M. 1995. *Diffusion of innovations.* New York: Free Press.

Rooney, C. 1993. Economics of pollution prevention: How waste reduction pays. *Pollution Prevention Review* 3:261–276, Summer.

Schmatz, E. 1997. Chrysler Corporation's Paint Pay-as-Built Program. Paper presented at Chemical Strategies Partnership Workshop, Chemical Strategies Partnership, in San Francisco, 20 November.

Tully, S. 1995. Purchasing's new muscle. *Fortune,* 20 February, 75–83.

Votta, T.J. 1998. Implementation of chemical management at Nortel semiconductors. Paper presented at Chemical Strategies Partnership Workshop, Chemical Strategies Partnership, in San Francisco, 21 May.

Votta, T.J., R. Broe, J.K. Johnson, and A.L. White. 1998. "Using environmental accounting to green chemical supplier contracts. *Pollution Prevention Review* 8: 67–78, Spring.

Wiseman, T.R., and B.J. Knight. 1996. Improving processes and partnerships. *Environmental Protection*, October, 30–34.

Williams, T.A., S.S. Brewer, P.N. Mishra, and O.W. Underwood. 1995. Pollution prevention at General Motors. In *Industrial pollution prevention handbook*, edited by H.M. Freeman. New York: McGraw-Hill.

INDEX